The Power of eCommunication

A comprehensive guide for using e-mail, voice mail, instant messaging, conference calls, and Web conferences

Laurie K. Benson

HRD Press, Inc. • Amherst • Massachusetts

HRD Press, Inc.
22 Amherst Road
Amherst, MA 01002-9709
800-822-2801 (U.S. and Canada)
413-253-3488
413-253-3490 (Fax)
www.hrdpress.com

ISBN 0-87425-756-5

Cover design by Eileen Klockars
Editorial services by Suzanne Bay and Sally Farnham
Typeset by Pracharak Technologies (P) Ltd., Madras, India

Printed In Canada

Contents

Introduction

Applications

Conclusion

Dedication

To the men in my life: My husband, John, and my son, Conner. Conner, it will be many years before you need to use a book like this, but e-communication is here to stay. I hope you'll find this book helpful years from now as *you* enter the world of virtual communication.

John, thank you for being my best friend, for *always* being there for me, for your never-ending love and support, and for knowing me better than anyone else on earth.

Mom, thanks for your endless words of encouragement and for making me believe that if I worked hard enough, I could be successful at whatever I chose to be.

Acknowledgments

To the staff at HRD Press—Chris, thanks for having faith and confidence in me from the very beginning. Bob, thanks for putting up with me wanting to change everything up to the last minute . . . and still making changes.

Thanks to my colleagues and friends, who patiently waited for me to finish this book and provided their perspective along the way.

Introduction

Technology and Today's Changing Business Environment

In today's environment, time is critical, and business professionals need to be productive no matter where they are: in the office, on a plane, or in a hotel room in another city, state, or country. Technology has come to the rescue! Or has it?

A debatable point, but one thing is certain: The same technology that businesses rely on today has forever changed how we communicate. It has given us the ability to conduct business transactions with people all around the world, often without ever meeting face-to-face or having a "live" conversation.

E-mail, instant messaging, voice mail, conference calls, Web conferences, and videoconferences have revolutionized how we communicate in corporate America and around the globe. In today's fast-paced business environment, companies are dependent on technology to link personnel across cities, states, and countries.

As an executive at one of the country's largest international consulting firms, I receive on a daily basis over 100 electronic mail (e-mail) messages, 10 voice messages, and countless instant messages, and I participate in and/or lead four to six conference calls a day. These communications are with people from around the world—people I will most likely never meet. Call that crazy? I call it normal communication in today's fast-paced, virtual world.

We are a nation dependent on these technologies. Virtual teams of people who are not co-located but who share responsibility for project deliverables and results are becoming a more important part of our business environment, and it is technology that makes this possible. Unfortunately, very few companies provide formal training on how to effectively manage communication through these resources. The purpose of this book is to offer some guidelines and things to consider regarding how to communicate more effectively through e-mail, instant messaging, voice mail, conference calls, Web conferences, and videoconferencing.

In this book, I'll provide you with tips and checklists that will help you create more-effective e-mail and voice mail messages and get the most out of your conference calls, Web conferences, and videoconferences. Whether you are a participant or the facilitator, the information will help you communicate more effectively. I'll also provide some ideas on how you can use these communication tools to help you build, maintain, or participate on a virtual team.

The way you communicate says a lot about who you are. It is also extremely important to your career. So, before you send your next e-mail, instant message, or voice-mail message, and before you facilitate or participate in your next conference call, Web conference, or videoconference, read these guidelines and think about what you really want your audience to take away from the messages you are about to send.

1 Applications

Communicating Effectively through E-mail

> Life is different: Where once Americans depended on the vagaries of the post office to communicate in personalized, written messages, now we send 2.2 billion e-mail messages a day, compared with just 293 million pieces of first-class mail (*U.S. News & World Report*, March 22, 1999).

In days past, we had the benefit of being able to communicate face-to-face. Businesses now operate globally, moving at e-speed; more often than not, we don't have the opportunity to meet most of the people with whom we communicate. What's more, most of our communicating in the next decade will not be done through formal letters or memos, but most likely through e-mail.

The individual with whom we are communicating might be a team member, our boss, our boss's boss, the head of our company, or a client. It is extremely important that we communicate clearly. Remember that the individuals receiving your e-mail will not have the benefit of hearing the tone in your voice or seeing your facial expressions, and might not be familiar with your communication style. Therefore, what they read is what they get . . . in black and white.

In many companies, e-mail is the communication lifeline. An e-mail message can be generated to an audience of one or an audience of a thousand in a matter of minutes—literally. Should that change how we communicate? *Definitely.* There are also legal issues that should be considered, as you will see in Chapter 7. A few years ago, all we had to do was type a memo and hand it to the appropriate individual, or mail it. That same memo can now be sent to an individual electronically, but it can also be forwarded to thousands within minutes.

Today's organizational structures are fairly complicated, adding an additional layer of complexity to our communications because we do not truly "know" our audience: We haven't had a chance to build relationships or get to know one another. Sometimes we're not even sure where an individual with whom we are about to communicate fits in within the organization.

Carefully consider the various communication styles and diverse backgrounds that make up today's organizations before you communicate a message: How often do we receive messages where the objective of the message is unclear? Even after reading the message, we ask ourselves, *What is the messenger trying to say? What is his point? Is action required by me?*

E-mail is easy to access and easy to use, and it is a fast way to send and receive information. It also has an informal feel to it, so people tend not to spend the same amount of time formulating e-mail messages or responses as they would with a more-formal method of communication.

The list of precautions provided below challenges you to take a good look at how you communicate through e-mail—not only through your use of words, but also through your use of fonts, graphics, and punctuation. You want your e-mail messages read and responded to in the time frame that you need, and these suggestions will help.

The most important thing to keep in mind when composing an e-mail is that all your reader has to go on is what's in front of her in writing—your e-mail message. Your reader cannot see your body language or hear your tone of voice, and might not even be familiar with the topic that you are writing about. Therefore, it is *critical* that you keep communication clear, concise, and focused on the facts.

Here is a list of things to remember and consider the next time you create your next e-mail, and before you send it.

- The overall guiding principle is to "think" before sending an e-mail. Be sure that you know what you want to say and what you want to achieve by sending the message.
- Make sure that e-mail is the appropriate vehicle for your message. Consider the objective of your message. A phone call or conference call might be more appropriate or get better results. In Chapter 6, you'll find guidelines that will help you with this decision.
- If the recipients of your message are not likely to know who you are, identify yourself at the beginning of your message or contact them first by phone or voice mail to introduce yourself and explain what you need.

  To make it easier for your recipient to place who you are and understand the role you play within your organization, you should include a "footer" or a "header," or some other type of signature line that clearly states your role within the organization, as well as a phone number where you can be reached. (Country codes can sometimes be a challenge to locate, so be sure to include it for your international colleagues.) For e-mail applications that do not have this capability, include it at the bottom of your e-mail. This will help the recipients better understand why you are sending them the message and give them easy access to a number where you can be reached. This might be more important in a large organization; in fact, some large organizations have defined specifics on what information they want the header to include. If you work for a large organization, check with your Communications or IT manager to find out if specific standards have been created.

At the beginning of the message, state the reason you are sending the message and what you want the reader to do.

- Here's an example of an e-mail message that accomplishes this:

From: Art Benson

To: Laurie Achgill

cc:

Subject: ABC Project Update

I am the project manager for the ABC Project. The purpose of this note is to provide you with a summary of the key findings from this project and recommend next steps to the committee. As you are the project sponsor, I would appreciate your feedback and perspective no later than this Wednesday, September 15.

Art Benson
XYZ Manufacturing Company
Director of Manufacturing
Direct dial: 1(555) 555-5555

Note: The footer in this memo includes Art's title within the company, but the note mentions the specific role and project that Art is working on. Therefore, it is still appropriate to include all of that information within the memo. This helps the project sponsor understand his role on the ABC Project within the context of his broader responsibilities.

- Do not deliver bad news or a harsh message through e-mail. In those cases, a phone call or person-to-person meeting is best.

  If ever in doubt, picture the person sitting across from you. Would you be able to say what you are about to say? If it is a difficult subject or message, a phone call or face-to-face meeting would be better.

- Do not say anything in an e-mail that you wouldn't say in person.

- Do not debate a topic through e-mail. Telephone the person to discuss your perspective and allow him to share his views. This will help resolve any conflicts and ensure that the desired outcome is achieved.

- When used effectively, e-mail can be a great tool if you need to share a new idea with colleagues. It's important to recognize that people absorb and respond to information differently. Send an e-mail message introducing the idea and outlining the benefits and objectives to give them the opportunity to review your message and then think about it before they respond. You can initiate a follow-up phone conversation to discuss it in more detail.

- If you receive negative feedback or sense frustration or confusion through e-mail, refrain from responding immediately with another e-mail. Instead, follow up with the individual on the telephone as soon as possible to discuss the feedback verbally. This will help minimize the potential of either party being misunderstood, and clear up any potential confusion.

- Know who your audience is and what they are interested in. Do not send messages to people who are not interested in what you are saying. They will be annoyed and might not read more-important messages from you later.
- Be certain that the individuals you include in the "to" field really are the individuals who need to read and/or act upon the information you are sending. As you think through the list, ask yourself why the recipient needs to see this message. Will he or she need to respond? Is this really an informational note? These questions will help you decide who should actually receive the message and who should be copied on the message.

  Generally speaking, individuals who are identified in the "to" field are the ones who need to know what you are saying and/or need to take action based on your message. Individuals who are listed in the "cc" field are receiving the message as an FYI (e.g., your boss and/or team members might want to receive a copy so that they know that you sent the message and what the message said). However, it is still important to let the receivers know in the beginning of the note what they need to do and why they are getting the message, especially if you think there will be some confusion. You might want to add a brief statement at the beginning of the note so that it draws the reader's attention immediately.

Note: From a communication perspective, it is important to know what your boss and/or team members want and need. I've worked with various bosses and teams who wanted to receive a copy of each communication that I sent out. Others didn't want to receive anything except a weekly or monthly status report. (Chapter 8 provides an example of a status report that you can send through e-mail.) Discuss this with your boss and/or team; solid communication is key to a successful team, but it's especially important in a virtual team situation. When you have a discussion around communication preferences with your boss and team, you might want to raise the idea of having individual and team status reports as a way to keep everyone informed of project status, key accomplishments, and issues.

Now, back to the main discussion at hand . . .

- Here is an example of an e-mail message that includes an introductory statement about who needs to do what with the message:

From: Conner Benson

To: Larry Jackson, Amy Thompson, Karen Gray, Clayton Robbins, Lynn Adams

cc: Jill Dunkin, Thomas L. Jones

Subject: Action Required—Project Team Update

Action required by October 1.

Please note: If you have been copied on this note, you are receiving this for informational purposes, and no action is required.

The purpose of this note is to request a status on the tasks that you are performing related to the ABC Manufacturing project. Attached to this note is a copy of the most recent workplan for the project. Please make the necessary updates and return the workplan to me no later than Friday, October 1.

Conner Benson
ABC Consulting Company
Project Director
Direct dial: 1(555) 555-5555

- Use the subject line for its intended use. This will provide the reader with a good idea of what the note is about so that he/she can decide if it needs to be opened now or can be opened later.
  - For example, if you need the reader to review the content of the note and respond, the subject field might read, "For your review and response." If the note is merely informational, the subject line might read "For your information/No action required—ABC Project Update."
- Strong opinions should be left out. Messages should focus on facts. Leave the opinions for a phone conversation or a meeting, when you have a better opportunity to clearly express your ideas.
- Keep emotion out of your messages. If your message is about a topic that arouses emotion or contains political, complex, or confidential matters, it is probably best not to communicate the message through e-mail.

  Having said all that, if you decide that you must send a message through e-mail that includes strong opinions or emotions, create the e-mail, save it as a draft, and re-read it later in the day (or sleep on it). Have a trusted friend, mentor, or co-worker read it. You might be glad you did. If after all these simple steps have been taken you *still* feel comfortable sending the message, then send it.

  You should also try to remain as positive as possible in all your communications (verbal or written). If you are having difficulty being positive, at least try to remain neutral.

  Again, a "live" conversation is recommended when strong opinions and/or emotions surround a topic.

- Humor is acceptable if it is in good taste and if you feel your audience will accept it. Sarcasm, however, is not appropriate, since it might be hard to detect sarcasm in an e-mail. Your comment might also be taken literally.

- If you are sending a lengthy or detailed message, I recommend that you provide:

  - An **estimate** of how much time it might take the recipient to act on the message. Otherwise, he/she might skip over it, assuming that it will take too much time.

    You can include this in the beginning of the message. If you are sending your message to executive-level personnel and you need them to take action, you might want to use the subject line to capture this information. For example, the subject line might read, "For your review and response. Approximately 10 minutes of your time is required."
  - An **overall summary** of the contents of the message at the beginning of the note. Within the summary, answer the basic questions of what, when, where, why, and how (or, at least, all that apply).
  - **Additional details** in organized sections. Some e-mail applications offer a function that allows the sender to put text or documents into collapsible sections. If the application that you are using does not allow you to do this, organize your e-mail in this same fashion using an outline format.

    This function is extremely valuable. It enables the sender to provide an overview at the beginning of the message and then attach details in sections in case the reader wants more details. This will help you organize your thoughts as you pull the message together, and it will help the recipient in reviewing the information.

Let's illustrate the points listed above. Let's say that you are kicking off a new project. This project involves a virtual team made up of individuals who are not co-located, but they share responsibility for project deliverables and results. You've decided to conduct a conference call with all the team members. (Good idea!) Prior to the call, you need to pull together information that they will need to get started.

If the team will be working virtually, it is even more important to have a clearly defined, well-organized plan of action: an objective statement, a workplan, expected outcomes, etc. Since this is a lot of information to communicate through e-mail, it is recommended that you put it into organized sections.

- Here is an example of an e-mail that has been developed and organized using the points above:

From: Eleanor Hall

To: XYZ Project Team

cc:

Subject: Action Required—Project Kick-Off Conference Call

Welcome to the XYZ project team! I am the project manager, and I am looking forward to working with each of you over the next few months. The purpose of this memo is to set up our first team conference call for February 15th, from 2:00–3:00 p.m. CST. Please plan to spend at least 30 minutes prior to the call reviewing the materials provided below.

The objective of this upcoming conference call is to officially kick off the XYZ project. More specifically, the objective is

- to introduce the team members and discuss how we will work together as a virtual team
- to discuss objectives and expected outcomes of the project
- to walk through the workplan and timeline
- to assign responsibilities and discuss next steps

continued

> This call is extremely important, as it will ensure that we have an overall direction and are working toward the same end result. Please make every effort to participate. If you are unavailable for the call, please let me know as soon as possible.
>
> Attached in the sections below are an agenda for our call, an objective statement and expected outcomes for the project, and a detailed workplan (which includes timelines, responsibilities, and next steps). I look forward to talking with each of you on the 15th.
>
> *Agenda for February 15 Conference Call*
> *Objective Statement/Expected Outcomes for XYZ Project*
> *Workplan/Timelines/Responsibilities*
>
> **Eleanor Hall**
> **ABC Consulting Company**
> **Manufacturing Project Manager**
> **Direct dial: 1(555) 555-5555**

- Proofread. Always check for misspellings and grammatical errors. Careless errors tend to imply that the sender doesn't care.
- When copying text from another application, proofread to make sure that all of the punctuation was transferred. Some applications will change the font and/or delete punctuation when copied into an e-mail.
- If you are sending an e-mail with extremely long attachments, remember that some individuals might be receiving your message remotely. Such e-mail messages are sometimes difficult to access. If you must attach a lot of text, call or leave the individual a voice-mail alert a day or two in advance so that they can expect it.
- If you are sharing information about a database or a Web site, provide a direct link to the database, in addition to the database path name (in case the link doesn't work).

- Keep in mind that a lot of different fonts, colors, and pictures increase the size of the message. It will take more time to receive or open these messages. Therefore, use these elements only when necessary, not just to make an e-mail look "pretty."
  - For example, if you need someone to focus on specific points within the e-mail, you might highlight them in a different color or font to make it easier for the reader to find or to announce a promotion or another type of celebration. A celebration e-mail would, of course, be an exception; you might want the message to be colorful to reflect celebration.

    Note: Color, font, and graphics don't always translate from one e-mail application to another. However, if the recipient of your e-mail has the same e-mail application, there should not be a problem.
- Mark the message "urgent" or "action required" only when necessary. Sometimes this function is overused, lessening the importance of a truly urgent message.
- Use the return-receipt notification function if you want to be certain that the individual has received your message. (Only some e-mail packages offer this capability.)
- Do not use aggressive punctuation (such as exclamation points) or bright colors or bolding unless you are certain that the emotion demonstrated through one of these options will not be misunderstood.
- Do not use ALL CAPS. Many readers might interpret this as being yelled at.
- Do not use slang terms. (Also, be especially careful when communicating with someone in or from another country.)

- Do not use abbreviations or acronyms unless the audience knows what they mean, and don't overuse them. Also, be sure to spell out any abbreviations or acronyms the first time you use them.

- When communicating with someone from a culture different from your own, be sure you are familiar with the protocol for that country. For example, in the United States we are very quick to call people by their first name. However, this is considered inappropriate in other countries.

- *Never* use offensive language.

- Do not send chain letters to business colleagues. It might lessen the importance of other messages you send to them later. Additionally, it might be against company policy.

- Never use your business e-mail address to send messages with offensive or pornographic content. Most companies have a policy stating that this behavior will result in immediate termination.

- Refer to your company policy before using your business computer or e-mail address to access the Internet.

  Some companies have a policy stating which Internet sites you can access and which ones to stay away from. Remember: If you use your company's Internet access, today's sophisticated technology makes it possible for Technical Support to identify the Internet sites you access.

- Refrain from downloading ".exe" files from the Internet. The file could be infected with a virus. If you have questions or concerns, always contact Technical Support.

- Be aware of virus trends. If you receive a file that you are concerned about, do not launch it. Contact Technical Support. Some viruses can be extremely dangerous because they erase data and programs.

- Stay current on the fight against viruses. Make sure you regularly scan for viruses. Check with Technical Support to make sure you have the most recent version of the anti-virus program that your company uses.

- It's important to remember that some people will print your e-mail message. Messages print differently, depending on computer screen width, printer type, etc. Use rulers in the application so that the message will print properly.

- Remember to save a copy of the message in your mailbox before you send it, if you think you will want to reference it at a later time.

- Prior to sending the message, verify that you have identified the appropriate individuals for the message and that the name in the "to" field is accurate.

  A colleague recently typed in the wrong e-mail address and sent her memo to a distribution list of more than 200 people. She was extremely embarrassed. Once a message is sent, you might not be able to retrieve it. You can only send an apology e-mail requesting that those individuals disregard your message.

- If you have the need to send standard e-mails to individuals on a regular basis (e.g., perhaps you are in a customer support role and send surveys, etc.), you should consider using a stationery template (if your e-mail application provides that capability). Perhaps the easiest thing to compare this to is form letters; it's the same principle. The difference is that the "letters" are really e-mail messages.

Some applications have this function; it allows you to save a memo into a template and use it whenever you need it, rather than re-create a new memo each time. Here are two examples of internal messages: one has the stationery "blank" and the other has the specifics.

From: John Benson

To:

cc:

Subject: Received Your Request

I have received your request for *(fill in the blank)*. I will begin processing that request immediately and will contact you if I need additional information. You can expect to receive a response from me no later than *(fill in the date)*. In the meantime, if you have any questions, please contact me.

John Benson
XYZ Appliances
Customer Service Department Manager
Direct dial: 1(555) 555-5555

From: John Benson

To: Dawn Johnson

cc:

Subject: Received Your Request

I have received your request for the report regarding how many repair requests have been made over the past 90 days. I will begin processing that request immediately and will contact you if I need additional information. You can expect to receive a response from me no later than the end of the day on January 17. In the meantime, if you have any questions, please contact me.

John Benson
XYZ Appliances
Customer Service Department Manager
Direct dial: 1(555) 555-5555

- If someone has sent you three separate e-mail messages with three different topics, consider sending only one message back, responding to all three messages.
- When forwarding a "string" of e-mail messages, summarize the messages as much as possible, and outline what you need from the reader. Don't expect that he or she will read all the messages. There might not even be a need to attach the "string." If there isn't a need, do not attach the others. Sending lengthy messages can frustrate recipients and also bog down your server.
- Do not use e-mail as a way to avoid talking with someone. Use it to be more efficient.
- Speaking of being more efficient, there is an instant message (IM) option available through e-mail services. If you have this, it's great! But don't overuse it.

  If you don't have IM, you might consider getting it for your team, because it's a great way to stay in touch. Refer to Chapter 2 for more information on instant messaging.
- Keep in mind that e-mail can be reviewed by others within your organization without your knowing it—and it is legal. Information relayed through e-mail is considered company property.
- It's important to keep in mind that e-mail messages can be forwarded to others. Be careful not to say anything that you wouldn't want others to see!
- If you are unsure about saying something, *don't say it.*
- Remember: Some things are better said or discussed than written. E-mails can be incriminating. The information below should help you keep that in mind.

An e-mail is hard to get rid of. It's almost impossible to take back what you have written once it has been sent.

E-mail replicates itself like bacteria in a petri dish. Once a message has been sent, it is virtually impossible to ensure that no copies survive . . . computer files are persistent by design. They tend to be copied and propagated and every corporation backs up disks. The chances are, a year after sending a message, there will be 50 copies (*Management Today*, February 1999).

E-mail messages can now be used as evidence in a lawsuit.

Electronic communication has added a new dimension to the discovery process through which lawyers search for evidence for use in trying lawsuits (*Association Management*, May 1999).

In 1970, the Federal Rules of Civil Procedure was updated to reflect that the term "documents" also refers to electronic documents.

Messages communicated, and possibly preserved, may contain evidence for future lawsuits stemming from discrimination, sexual harassment, price fixing and product or other liability claims (*Business First*, July 1996).

Suppose you e-mail an employee whose project is running late: "I don't want to hear about how it isn't working right. I just want you to get the thing done." Is that blunt talk from a tough boss who doesn't want excuses? Or, is it an instruction to cut corners? You can bet which way a[n] . . . attorney would interpret it (*Forbes*, July 1998).

- It is extremely important that you become familiar with your company's e-mail policy. After reading the text above, you probably have a better appreciation for this point.

  Your company might not have a policy. Analyst Jim Browning of the Gartner Group estimated in 1999 that half of the leading U.S. companies do not have e-mail policies (*Management Today,* February 1999). If your company has one, it might provide guidance on what is considered appropriate and what is not.

- If you want to send a confidential e-mail message that cannot be forwarded or printed, some e-mail applications provide delivery options that allow you to encrypt a message and/or prevent copying or forwarding of the message. Individuals can often find ways around this, so using these options won't provide a fool-proof solution. Server administrators will still have access to these confidential messages.

- Treat the information within an e-mail as copyrighted material. In other words, do not use the work that someone else put together without giving them credit.

  If you want another member on your team or your boss to see the information sent to you within a note, actually forward the original message and copy the author on the note if you feel that's appropriate.

  This not only helps to build trust in our virtual environment, it also provides the name of the original source in case the recipient of your note has questions about its content.

- If you intend to be away for an extended period of time (e.g., on vacation), send an e-mail message to your co-workers, boss, and others you have been working closely with to let them know that you will be out and when you will return. Be sure to include the name and phone number of the individual "filling in" for you in your absence. Another option would be to create an "extended absence" setting. Some e-mail applications allow you to create an "extended absence" form to automatically send an e-mail message to anyone who sends you an e-mail message, letting the sender know that you are away and when you will return. This helps manage the expectations of the sender as to when you will receive and reply to her message. She might find that she needs to send the message to someone else in your absence.

MANAGING THE MASSES: *How do you deal with all that e-mail?*

Now that you've been given a few hints on how to communicate more effectively through e-mail, let's look at how you can manage the daily mass of e-mail communication. My experience is that we can spend literally all day reading and responding to e-mail if we let ourselves. However, you must remember that you still have a job to do. E-mail is merely a tool to help you do your job. Do not let it rule your agenda for the day. Here are some things that you might consider:

- Most e-mail applications have a way to alert you when you receive new mail. Sometimes it's a "ding" or some other sound. Turn it off or turn the volume off. If you don't, it will distract you. Some people feel that as soon as they get an e-mail, they have to open it right away. Remember, if someone needs you that urgently, they can always pick up the phone and call you.
- If turning the volume off doesn't help, then work from a local copy of your mailbox, and only copy new messages to your local mailbox at scheduled times throughout the day. If you don't know how to do this, contact someone in Technical Support.

  I have some colleagues who turned off the "ding," but when they receive a new e-mail, they see a new message because it is the color red or there will be a written message saying, "You have new mail." Well, they still feel the need to open it right away. Working from a local replica copy of your mailbox and scheduling replication times will allow you to still have access to the messages that you need, but it will also save you from the constant distraction of receiving new mail.

Think of it this way: When the U.S. Postal Service delivers the mail, it sits there until you retrieve it and open it. It's the same principle. If you are urgently waiting for a particular message, you can go back to the server copy of your mailbox on occasion and check for mail received by a specific person. This will help you be in more control of your e-mail.

- Schedule set times throughout the day to review and respond to your new e-mail messages.

  Consider scheduling an hour first thing each morning, 30 minutes around lunch time, and an hour before you leave the office. The amount of time you set aside and the number of times a day that you check your messages really depends on how many e-mail messages you receive on a daily basis, and generally how urgent they are.

- Consider categorizing your e-mail into "folders" as soon as you get it.

  If you're like me, you need to have the trash can close by when you're going through the mail delivered by the mail carrier. I sort the mail into stacks according to the action required (e.g., bills, informational, to follow up on, etc.). The rest goes in the trash. You might put a similar method in place to help you organize your e-mail. You can categorize the mail by action required and date due. Once you've responded, you can move it to a category by topic. It sounds difficult and time consuming, but it's not. If you want to get a handle on your e-mail and want to plan your day, this is probably the best way. It really doesn't take long at all to take these extra steps, and it makes a world of difference.

- Delete junk e-mail messages immediately. I get approximately 50 of these a day. I delete them first. It's a mental trick: It helps me feel like I've accomplished something, and my mailbox seems smaller because there are fewer messages that I need to respond to.

  One exception is "humor" mail. It might be a good idea to save some of the humor stuff for later on, when you really need it. A friend of mine recommended this to me. She said that when she receives messages that include humor or an uplifting thought, she categorizes them as such. When she is having a tough day, she goes to that category and re-reads some of the messages. This lightens her mood and lifts her spirits.

- As soon as you reply to a message, delete the original message if you will not need to reference it again.

- Always be sure to delete messages that you no longer need to save. To help with this, you can:

  - Make sure that your computer is NOT set up to automatically save a copy of each mail message that you send. This will require that you actually think about whether or not you need to save a message, rather than it automatically happening for you.

  - Delete a message as soon as you've read it, if you are certain you will not have to go back to it again.

Mailbox maintenance can be a real time consumer if you don't take care of it along the way. To maintain your mailbox, delete old files, archive old files, and categorize notes so that you can find them more easily at a later time. If you follow these steps, mailbox maintenance will not be such a challenge.

This was a big lesson-learned for me. For several months, I was extremely busy at work. I didn't really try to stay on top of my e-mail by deleting messages that I no longer needed or categorizing them. Well, my computer crashed. Once Technical Support got it working again, we found the problem. I had no memory left on my hard drive! They told me that it was critical that I immediately delete old messages. It took me two full days and some weekend time to delete and categorize my mail.

I also used to procrastinate in maintaining my mailbox. I'd get a telephone call every couple of months from Technical Support telling me that my mailbox was too big, that I was *hogging* the server, and that I needed to delete messages immediately. Now I keep up with my e-mail maintenance and feel more in control. (I don't get those phone calls anymore, either!)

In this chapter, we've discussed how you can communicate more effectively through e-mail, and we've provided some tips to manage all the messages you receive. We are all looking for ways to manage communication in today's fast-paced world. Let's not let e-mail control our day and how we do business. Electronic mail is an extremely valuable communication tool. Nevertheless, most executives feel that they spend too much time reading and responding to e-mail messages. Each day the number of e-mail messages we receive and send increases, and will only get worse as more and more people get online. Before you send another e-mail message, make sure that you are using the right communication vehicle and that your objective for the message is clear. That will make the time spent far more efficient and effective.

2

Using Instant Messaging to Add Value to Business

Instant messaging (IM) is a technology people have been using in their personal lives for quite some time now, but business is adopting it to improve communication and productivity. It offers cheap, efficient, real-time communication capabilities.

> Instant Message (IM) . . . some people hate it and some people can't live without it. There are now an estimated 250 million users of Instant Message *(Microsoft PressPass,* May 27, 2003).

IM brings new capabilities that can really help businesses improve communication and even manage team members that are geographically dispersed, *as long as it is used effectively*. There are clearly advantages to using IM in business.

IM can help an organization improve communication and productivity, as well as improve "connections" between virtual teams, but the technology also has a downside. A few examples of advantages and disadvantages of instant messaging are explained below.

Improves Visibility

IM allows supervisors and employees to see who is logged on their computer (as long as they are logged on to IM). This can help supervisors monitor the work hours of team members, especially when the team is spread out over several locations.

Increases Accessibility to Team Members

In the past, if you had a question or needed information from a team member, you would have to send an e-mail or call the individual. Sometimes people are not even accessible by phone. With IM, a team member can still see the instant message and possibly be able to respond immediately. This is why IM is called "real-time" communication. If you are on a conference call, try NOT to respond to instant messages; instead, use the "away" feature built in to most IM applications. It will send an automated message back to the sender, letting them know that you cannot respond right away. I recommend creating a specific "away" message that you can use when you are on a conference call. It might say, "I am on a conference call that requires my full attention. I will respond to you as soon as the call is complete." Now, if you want people to be able to reach you immediately (if the need is urgent, for example), you can add your cell phone number at the end of the "away" message and indicate that if the matter is urgent, you can be reached on your cell phone.

Reduces Communication Costs

IM can also help your company reduce communication costs. Rather than picking up the phone and incurring long distance charges, team members can communicate directly, with the same results.

Okay . . . we've talked about some of the benefits of IM. Let's look at the challenges associated with IM.

Is Not Reliable

IM is not 100 percent reliable. I can't tell you how many times I've gotten "knocked off" IM. You can miss a message this way and be completely unaware that you've missed anything. That's why it's important to give people time to respond to your instant message. Send them another instant message to see if they got your previous instant message. Never assume that they just don't want to respond; there is a good chance that they didn't get your instant message because they were disconnected and had to log back on.

Let me give you another example of why an instant message might not be received: If someone is typing and hits the "escape" key at the exact same time you send them an instant message, they can "escape" right out of it. That has happened to me many times.

Can be Distracting

Some people say that IM makes it a bit too easy to communicate with people, and that it can distract them from potentially more-important tasks. It also tends to create a sense of urgency.

Provides No Official Record of Communication

IM is easy to use, but it should not be used if you need to document conversations. For example, if you are a customer service representative and your customer contacts you over IM, there is no official way to keep a record of that conversation. However, there is a way to set up your computer to save your IM exchanges, so be careful. You never know who might read them.

There are some complications with IM, but the benefits outweigh them.

Here are just a few suggestions on how you can use IM to improve communication within your team.

Use IM with Conference Calls and Web Conferencing

IM can be used in conjunction with conference calls and Web conferencing to:

- Provide coaching, answers, or recommendations to a presenter/speaker (assuming that both of you are logged on to IM). *Note:* Be sure to turn off the volume on your computer. Otherwise, participants on the call will "hear" that an IM exchange is happening.
- Provide presenters with background information about a question being asked.
- Allow presenters to correspond, share ideas, decide who will answer questions as they are asked, etc.
- Allow the facilitator or presenter to seek informal feedback from participants.
- Allow you to play a game with your team. For example, you can tell team members that the first person

who sends you an instant message with the correct answer to a question you pose will win a prize. This game is a lot of fun, and it can really help bring a virtual team together.

Use IM to Stay in Touch with Your Team

IM can also be used to stay in touch with your team. While it's not the same as being there, it can provide a much-appreciated way to connect.

- During a stay-in-touch project, send an instant message with some encouraging words to your team members.
- Send a "smiley face" to a team member, perhaps with an instant message that says, "Here's a smile in case you need one today."

Expect many additional enhancements to be made to this communication tool over the years: IM is here to stay.

3

Communicating Effectively through Voice Mail

Most companies use voice mail (as they do e-mail) as a primary means of communication. Effective use of this tool is even more critical because companies work globally and across several time zones. Voice mail enables you to send and receive messages remotely from any phone in the world at any time, crossing time zones. Using a combination of e-mail and voice mail, an executive can go several days without talking voice-to-voice with anyone, yet is still able to get the job done and get it done quickly, regardless of what time it is!

Voice mail is a more informal form of communication. Like e-mail communication, voice mail allows us to share details, brainstorm ideas, and express opinions. However, voice mail is a more personal, informal communication because the receiver hears your voice and specific tones.

It is only a one-way conversation, but it lets you offer explanations, offer analogies, and basically just talk through something that might be difficult or confusing to put into written text. Voice mail can also offer a more immediate response, particularly if the individual is traveling and does not have access to his or her e-mail. It sometimes takes less effort for the person receiving your message to respond verbally than to write out a text response, which is another plus.

The voice mail form of communication is informal, but it is still a tool meant for business communication. You still have to communicate clearly and professionally, so learn how to use this tool effectively.

Here are a few things to consider when creating your next voice mail message:

- Determine if voice mail is the right way to deliver your message. It offers only one-way communication, so be mindful that your message won't leave a lot of questions unanswered. Ask yourself, *Will my message just prompt an exchange of additional voice mail messages? Will it be similar to phone tag–only "voice mail tag"? Maybe it would be easier to just pick up the telephone and have a "live" conversation.* In Chapter 6, we'll outline other things you should consider as you decide whether or not voice mail is the right vehicle for your communication.
- Speak confidently and clearly. Put your best voice forward, but be natural–as natural as you would be if you were speaking with someone in person or making a presentation. Add some of your personality through voice inflection and laughter, if it is appropriate. Again, just speak as you would in a presentation, but be careful: During a presentation, participants can see you. With voice mail, the audience will not be able to read your body language.
- Be sure to provide your name, contact number, and the organization you represent. (If you are sending the

message to personnel within your organization, you can just state your role within the organization, unless the individuals receiving your message know who you are and how to reach you.)

I have received numerous messages from individuals who did not leave a contact number and subsequently had to spend hours searching for that information.

- Here is an example of a voice mail message that provides all the necessary information:

> "This is Sandy Johnson, and I am the district sales manager. The purpose of this message is to make you aware of a meeting that is planned for next Thursday, the 23rd, at 3 p.m. You will receive more details through a follow-up e-mail message within the next couple of days. Please mark your calendars and plan to participate in this meeting. If you have any questions, I can be reached at (555) 555-5555."

- Keep your message as brief and to-the-point as possible, but be sure to put the objective of the message at the beginning.
- If you are contacting an individual within your organization and you get her voice mail, consider hanging up and leaving the message on your company's internal system. This will allow her to send a response voice mail without having to look up your number.
- If you are sending a very detailed message:
  - Make an outline for yourself to help guide you when you are recording the message.
  - Manage expectations about the length of the message at the beginning of the message. Specifically, state that the message is long and that it might require two or more calls in order to cover all of the points. (Try to keep it to just one message.)

- State how many key points the message will include.
- State the importance of listening all the way through (depending on the topic and audience).
- Keep the outline and key phrases in the message, in case there are questions that come up. You might even want to send a copy of the message to yourself for future reference.
- Prior to sending the message, listen to it to make sure it is clear. Be sure the content is accurate and the tone is appropriate.
- Here is an example of a voice mail message sent from a regional sales manager to his sales staff:

> "Hi. This is John Benson, with an important message to all Southeast region sales personnel.
>
> Let me start by setting your expectations about this message. This message will be less fun and more lengthy than what you are used to receiving from me. It is important that you listen to all of it. Please do so.
>
> The purpose of this message is three-fold. First, I want to comment on the excitement generated by our new president, Stanley Smith, and the results of our recent sales meeting. Second, I want to explain the reality of our current sales objectives and how that impacts how we operate. Third, I want to ask for your help."

More details of the message will follow. As you can see, John outlined his voice mail message much like he would a memo, with an introduction and an overview of the key points the message will cover.

- If you are recording a message that needs to go to a large number of mailboxes, you might want to consider recording it, sending it to yourself, and then forwarding the recorded message. This will minimize the risk that you will be interrupted or miss someone's number and hence fail to include them in the distribution.

If something like that does happen, you will have saved a copy in your mailbox, which you can forward later.

It took me a while to learn this lesson. I started sending a copy to myself only after several experiences where I recorded a very long voice mail and either got interrupted while sending it out or pushed the wrong button and deleted the message that I recorded. Trust me, sending yourself a copy first is a habit you'll want to adopt. If you don't, the first message that you have to re-record will probably convince you of its importance. Of course, after you are certain that you no longer need that message, be sure to delete it so that your mailbox doesn't get too large.

If you frequently send messages to the same group of individuals, consider setting up a distribution list. This will allow you to send the message to just one member. It will be distributed using the numbers in that distribution list.

- Cellular phones are an integral part of doing business. If you have a quick voice mail to send, the background noise on a cellular phone is probably acceptable. However, do not attempt to leave a long, detailed message when using a cellular phone. Oftentimes, cell phones cut out or the background noise makes the message less clear. This can lead to the person on the receiving end of the message missing key components of the message or becoming frustrated because he has to listen to the message several times. This has happened to most of us, and it's extremely time-consuming and frustrating on the part of the sender and the receiver. The receiver has to listen over and over to the message, and ends up having to contact the sender anyway because the message was unclear.

  If you must use a cell phone, pay close attention to the clarity of the connection and whether or not the connection cuts in and out while you are recording your message. Replay the message to double-check the clarity before you send it.

- Listen to your recording to make sure you are satisfied with the message. If you are unhappy with the message, re-record or edit it. If you are not sure how to edit a message, contact your telecommunications manager or vendor.
- Always create voice mail messages with the knowledge that they might be forwarded to others. Do not say anything you feel uncomfortable with. Some voice mail systems will allow you to mark your message "private," but it's important to remember that the telecommunications manager can still access your message, if necessary.
- Do not send a voice mail about a topic that will raise a negative reaction or a great deal of emotion in your voice or cause controversy. Save those conversations for a telephone conversation or a face-to-face meeting.
- If you will be out of the office for an extended period of time, update your recorded message to inform callers and provide them with pertinent information and who they can reach in your absence.
  - Here is an example of an extended-absence greeting:

> "Hello, you've reached Melony Achgill. I am traveling out of the country this week and will have limited ability to check my voice mail. If you have an urgent need, please contact Dawn Johnson at (555) 555-5555. Otherwise, leave me a message and I will return the call as soon as I can."

- Do not use voice mail to deliver bad news or a harsh message.
- Do not use voice mail to avoid talking with someone.

- Do not use voice mail to avoid or resolve conflict. These things should be dealt with in person.
- Avoid repeating information. Say it once.
- If any action is required on the part of the recipient, state this early in the message, and be sure to note when action is to be completed.
  - Example:

> "This is Conner Benson, with a message for Emily May. The purpose of this message is to make you aware of a meeting and provide you with an overview of action required by you. The executive committee will hold a meeting on November 20th in New York. You will receive via e-mail the agenda, pre-work, and additional details for the meeting by Wednesday. Please plan to have all pre-work completed and submitted no later than November 18th. If you have any questions after you receive the e-mail, let me know as soon as possible."

- If the message is going to more than one person, state that you will provide the distribution list for informational purposes at the end of the message.
  - Example:

> "This is Luanne Eskew, with a message going out to all of the ABC Project team members. I will provide a list of names included in this distribution at the end of this message. The purpose of this message is to remind you that status reports are due no later than this Friday. If you have any questions or need to discuss this with me, I can be reached at (555) 555-5555. This message is being distributed to: John Smith, Timothy Johnson, Jeremy Faulhaber, Mary Benson, Tom Jones, Kristin Robbins, Kelly May, Jesse Timms, and Mark Achgill."

- If you are contacting someone directly and voice mail picks up, be sure to state your name clearly and provide your phone number, in case the receiving party has questions or needs to follow up.

- If you are forwarding a message from another person, offer a brief explanation in an introductory message.
- If a message has been forwarded multiple times and has several introductions in it, you will want to offer a summary of all the messages and state that there are several introductions. You can also delete all those introductions before forwarding. If you are unsure how to do this, read the instructions for your voice mail system, or contact the telecommunications manager in your office.
- If you are waiting for a message from a specific individual (e.g., your boss) but you don't have time to go through all the messages waiting for you, search for the message from that individual. Most voice mail applications have this ability.
- *Never* use offensive language. Even if you think you know your audience, it's inappropriate and unprofessional.
- If you need an immediate response, don't be afraid to mark the message "urgent." Just don't abuse the privilege.
- If you need to know when someone has received the message, be sure to attach a confirmation. Most voice mail packages offer this technology.
- If you will be away from a telephone and are not able to leave a message at a particular time, you can record a message in advance and queue it to be sent on a specific date and time. Most voice mail packages offer this technology.
- Delete messages as soon as possible. Saved messages can take up a lot of space on the voice message system.

If your voice mail runs out of space, you will not be able to receive additional messages. This is extremely frustrating for people trying to leave you a message.

- Always try to respond to your voice messages within a 24-hour time frame, at least. If you have to do research, let the caller know that you received his message and when he can expect an answer from you.

4

Getting the Most Out of Your Conference Calls

An increasing number of companies prefer to use teleconferencing rather than schedule face-to-face meetings to bring people together from around the city or around the world. Conference calls reduce or eliminate the need to incur heavy expenses related to travel and loss of staff time, and are becoming an effective way to manage costs.

Surveys indicate that business professionals spend an average of three hours a day in meetings, and that face-to-face meetings can cost up to seven times more than meetings conducted through conferencing technology.

A conference call is, in many ways, a virtual meeting. It offers the team and the organization a way to maintain communication and keep people productive. Conference calls are less expensive than meetings and allow participants to informally communicate more information than they can by e-mail. A conference call allows everyone to ask questions, provide input, and openly discuss issues in real-time.

Teleconferencing is a powerful communication tool. An executive I worked for often said, "It's amazing. All it takes is a couple of conference calls and some follow-up e-mails and/or voice mails, and we have turned the organization on a dime."

Like meetings, effective conference calls require strong facilitation skills and participation from all involved, as well as a basic knowledge of the subject at hand and a general understanding of the audience for the call. Conference calls offer tremendous opportunity to share knowledge, roll out new programs, and keep virtual teams working effectively.

In this chapter, I provide some tips on how to get the most out of every conference call, whether it is a one-time call involving 100 people you don't know, a conference call with company leadership, or a biweekly status call you have with your virtual team members.

Prepare for a conference call much as you would for a meeting. Here are a few things to consider when preparing for, conducting, and participating in a conference call.

Preparing for the Call

- First things first. Clearly document the objective of the conference call to make sure it is the appropriate forum for your message. Perhaps a meeting is necessary.
- Clearly document what you expect/want the outcome(s) of the call to be.
- Consider the length of the call. I do not recommend that a conference call go on for more than an hour and a half. After 90 minutes, participants are likely to lose their focus and need a break.

  If the topic warrants a longer call, you might consider breaking it into segments or offering a lunch break in between. If you offer a "break" in the call, you will need to discuss this with the conference call vendor to let them determine the best way to manage the dial-in and connection process.

One person might have to leave one line connected so that the connection is not terminated. On the other hand, if the break is long enough, the vendor might recommend that everyone just dial back in through the same number.

- Determine the audience for the call. Conference calls are really "virtual meetings." Consider what Steve Kaye writes in his book, *The Manager's Pocket Guide to Effective Meetings:* "Effective meetings include only those people who can make significant contributions to the process, which is usually fewer than a dozen. Marginal contributors, spectators, curiosity seekers, lost souls, and trouble makers should be left out. If they want to find out what happened, you can send them a copy of the minutes."

  One exception is when a conference call is used to roll out and discuss new policies. In that case, "spectators," "troublemakers," and "lost souls" will probably need to be invited, because everyone will need to be aware of a new policy.

  I would like to provide a contrasting perspective on the "troublemakers" comment. Kaye recommends that these individuals not be invited to your meeting, and many professionals would agree. I suggest that you reflect on why they are considered "troublemakers." Some individuals are labeled as "troublemakers" because they are vocal, act negative, or consistently offer up different ideas from those presented. (Whether we admit it or not, at some time or another, each of us has probably been labeled a "troublemaker.") Some "troublemakers" are just passionate about what they believe in, whether it is a policy, a process, or simply another team member.

The advice to "pay now or pay later" might well be applied with these individuals. Identify all your troublemakers or passionate possibilities, and talk with them individually about the ideas, issues, or projects you wish to present during your conference call. This will give them an opportunity to share their concerns and ideas with you, and they will feel that they have been listened to (this is usually what they are craving). They might indeed have some really good ideas or help you enhance your ideas or resolve issues. Additionally, this meeting with them will give you an opportunity to get their buy-in early in the process. Sometimes the best way to cure a "troublemaker" is to win him over. Of course, you'll be the only one to determine if it is appropriate, possible, or worthwhile in your specific situation.

If you get the buy-in of some of these folks, it might change the tone of a conference call or meeting for the better.

- Determine the number of participants for the call. If there are a significant number of individuals who need to participate (maybe you are rolling out a new training program to all executives), this might make the cost to conduct the conference call very expensive. In this situation, here are a couple of options you can consider to help reduce the cost of the call:
  - If there are several people who need to participate in the conference call, try to identify an individual within each location who can help arrange for an office or a conference room so that everyone can participate. You will need only one dial-in port (phone line), and this will save money.
  - Identify (or ask the appropriate people to identify) one to two individuals from the team to act as

"trainers"—people who can participate in the conference call and, acting as trainers, take the information learned back to their team. Consider having the call recorded and available for a predetermined period of time to assist these "trainers" in taking the information back to their team. You have to be careful, though. Before you decide to have the call recorded, you'll need to consider the following:

1. How will you be charged for the use of the recording? Will you be charged each time someone dials in to listen to the recorded call, or will it be a flat rate?

2. Will sensitive information be discussed during the conference call, making it inappropriate for recording?

- Determine how many presenters you will need and who they will be. If you are the primary facilitator and presenter, invite other knowledgeable speakers as appropriate or ask that someone else be the facilitator. It is very difficult to effectively facilitate a call *and* be the primary speaker. An exception to this is if the call will be without debate or difficulty (e.g., status meeting call, general update/information-sharing call).

- Determine the exact time and date of the call. While you are considering time options, be sure to factor in the various time zones for participants. If you can't seem to find a time that works for everyone, consider having the call recorded so that participants who missed the call can dial in at a later time.

- Consider scheduling a "preparation" conference call with the speakers to make sure that everyone is clear on expected outcomes and their role in the conference call.

- Decide whether or not you will provide handouts for the call. If so, consider the following:
  - Keep the documents simple and easy to follow.
  - As appropriate, add color (but not too much) to the presentation material. Some people are more visual; using some color will help them remain focused during a lengthy call.
  - Create a detailed and organized agenda for the call and distribute it to participants prior to the call.
  - Here is an example of a detailed agenda that a project manager might use to kick off a conference call with a new project team:

Topic and supporting documents	Speaker(s)
Introduction	
• Remind participants of the objective or reason for the conference call.	Project Manager (*insert name*)
• Introduce team members, and discuss how the team will work together. *(Consider attaching a profile document for each individual, as mentioned in Chapter 8.)*	Project Manager/All
Walk through the detailed workplan/ project time line. *(Attach workplan document)*	Project Manager
Assign responsibilities, and discuss next steps.	Project Manager
Open discussion/Question and Answer session.	All

- Use diagrams and presentations, as appropriate. Some people are visual thinkers who need a "picture" of what you are saying, especially if it is a complex or new idea.
- Be sure to provide your participants with enough time to receive and review the documents prior to the call.
- Some people might be receiving these documents while they are traveling. If the document is lengthy, they might have difficulty receiving it. This is another good reason to provide materials a couple of days in advance of the call. If the documents you need to send are long and they are critical to the success of the conference call, contact the individual who is traveling and ask if she/he wants you to send them electronically, by overnight mail, or by fax.

- Determine if you will demonstrate a database or other tool. If databases will be discussed or demonstrated, provide database links and database path names. Also, be sure to set aside extra time so that you can fully describe and explain the tool and answer questions.
- Consider using an electronic whiteboard if you intend to draw pictures/diagrams to explain something to your virtual audience. Electronic whiteboards are especially valuable when conducting a brainstorming session during a conference call. They allow you to draw your idea and quickly send it through a presentation to the participants via e-mail. If you have an electronic whiteboard and want to do this, you'll need to allow for extra time prior to the conference call to review the instructions and prepare yourself. You also need to be sure that it will work as planned. There are many kinds of whiteboards available.

  Electronic whiteboards are good for virtual presentations, but they are especially helpful when some participants are "live" and some are virtual, because you will end up drawing pictures or writing notes while you talk with the "live" participants that you will want to share with the virtual participants as well.

If you need more information on the capabilities of electronic whiteboards and vendors, there are several Web sites worth browsing through:

- www.interactivewhiteboards.com (Numonics Corporation)
- www.teamboard.com (Teamboard, Inc.)

One tool that works even better than standard electronic whiteboards is what I call "eMeetings," which offer the opportunity to share information and brainstorm in "real-time." All participants using computers can view the same information at the same time. They can also view changes as they are being made and see pictures as they are being drawn. Have you ever been on the phone with someone and said, "I wish I could just draw you a picture"? Well, your wish has come true: It's called *Web conferencing*.

Web conferencing makes it possible to deliver presentations online in a more collaborative fashion. You can present PowerPoint slides, draw pictures, review a Web site, play games, and do much more. We will discuss Web conferences in more detail in Chapter 5, but here are a few Web sites if you are interested now:

- www.messagebank.com
- Sametime, an application available through Lotus (www.lotus.com)
- NetMeeting, available through Microsoft (www.microsoft.com)

- When sending documents in advance, recognize that no matter how much time you give them, some people will not review all the documents. Be prepared to provide a high-level overview and then as much detail as the participants need. You want everyone on the same page, from the beginning. Doing an overview is much easier than assuming that everyone has read the material and finding out that they haven't, in which case your conference call won't be productive because some people won't be prepared.
- Determine which conference call service you will use, using the questions provided in the "Setting up the Call" section below.
- If the topic is complex and participants come from cross-functional groups, you might want to send material to a small group of people and ask for feedback prior to the call. This can help you better prepare for the call: You can determine whether or not the messages are clearly documented or the topic is just too complex for a conference call.
- Once you've determined that a conference call is in order, contact the telecommunications manager in your office to determine if there are preferred vendors, specific discounts, and/or specific procedures you should be aware of. If your company does not have a preferred vendor, contact your local operator for the name of a vendor, or search the Internet for a provider.

Setting up the Call

- Contact the vendor to make a reservation for the call. Reservations can usually be made from 30 minutes (or less in some cases) to one year in advance. You will want to set up the call as far in advance as possible to give yourself time to organize the call and communicate the call-in information to participants. (With 24/7 conference-call capabilities, this step might no longer be necessary. For more information on 24/7 conference call numbers, access www.messagebank.com.)

- If you need to have more than one conference call with the same audience or you intend to hold regularly scheduled calls, consider requesting the same dial-in number for each call. This makes it much easier for the participants. They'll probably begin to memorize the number.

  A word of caution about this recommendation: If you have several conference calls in a row (back-to-back) and are using the same dial-in number, be sure to allow at least 30 minutes between calls that use the same dial-in number.

  Consider the following key points when making the conference-call reservation:

  - Do you want entry/exit tone beeps? These beeps are helpful to the facilitator because they let him/her know when people are coming on or dropping off the call. However, if it's a call involving more than 25 people, the beeps can be distracting. There is no right or wrong decision here; it depends on your personal preference.

    To help you make that decision, consider:

    1. Who is the audience for the call?
    2. How many people will be on?
    3. How important is it that everyone be connected on the call *and* remain on?
    4. Would it be more appropriate to have the vendor provide you with a list of attendees and the exact times they entered and exited the call, to assist you with follow-up?

- Do you want to reserve a presenter's time allocation 15 to 20 minutes prior to the call? If the call is very large, with multiple presenters, consider having the presenters start preparing for the call and then conducting a group debrief on a line of their own so that they can prepare and debrief as appropriate. An operator will link the presenter line into the main call-in number to begin the call when she is instructed to do so, and will link the presenters out of the main number upon completion of the call.

- Do you want the call recorded? Most vendors will tape and transcribe a call for a fee. If the call has a fairly simple subject and few questions and answers (e.g., a status call with a team), this service is probably not necessary. However, if the call is with an audience you are not familiar with and covers a fairly complex topic likely to generate a lot of questions (e.g., a new policy or procedure), consider having it taped. This will allow you to focus on the call as the facilitator, rather than be concerned about taking notes as well.

Note: If you are ever the primary facilitator on a conference call that you didn't arrange to have recorded and a lot of complex questions come up that you think you will miss in your notes, you have two options: (1) Ask someone in the audience to take notes (if you have upper management on the call, you probably should not do this) or (2) Ask participants to hold tight for a moment while you ask the operator to begin recording the call so that you can focus on the topic at hand. Most participants will understand and will wait. If they continue with the discussion anyway and key points are being missed, your responsibility as primary facilitator is to remind them that this portion of the discussion might not appear in the minutes.

- For calls involving more than, say, 25 participants, consider placing the participants in "listen-only" mode. This will prevent people from talking over one another. Listen-only is probably most effective in an information-dissemination type of call (e.g., a rollout of a new policy or program).

  Note: Listen-only mode is a function that can be set up through the conferencing service. This function automatically puts all lines on mute except for the presenters' lines (presenters are identified ahead of time). This can help keep background noise minimal. However, if you have a small enough group, you can keep all lines open.

  I must admit that the decision to use "listen only" is a hard one to make. I've conducted many successful conference calls involving 50 or more people without using this function. One key advantage to not having this function activated is that participants feel more free to interject questions and comments throughout the call. Even if they choose not to speak, they feel less restricted.

  The listen-only function, however, still allows individuals to ask questions by using special features such as pressing star (*) and then 1 on their phone. If you decide to use the listen-only option, be sure to give participants verbal instructions on how to interject a question. (You should get these from your vendor.)

- What is the cancellation policy? Most vendors have a no-show policy that you will need to be aware of.

- What happens if you invite a certain number of people to participate (let's say 20) and they invite others, making the total 25? Will the vendor provide additional ports? Or will they not let the caller come on? Find out how many lines they will add automatically, before you decide how many lines to reserve.

- Ask what and how you will be charged for any additional or unused lines.
- Ask if the conference call company has a "helpful hints" document that you can review prior to the call.
- Consider whether or not you want a formal introduction of the call by an operator. If so, be sure to provide the conference call operator with a script as to how you want the conference call introduced, along with the names of all the presenters and other key information. You might also want to schedule a discussion with your operator a couple of minutes prior to the call to make sure everything is in order and that he understands your expectations.

Communicating the Call

- Be sure to send the initial communication inviting the participants to the conference call as soon as possible. This will allow participants to set aside the time on their calendar. You can send a communication through e-mail or voice mail: Simply state the time, date, and objective of the call, and follow up with more details as the date of the call gets closer.

  In the follow-up communication, be sure to note that the call will be recorded so that participants can dial in at a later time (if that is the case).
- Be sure that you have invited the appropriate speakers and participants to the conference call. Clearly state the objective in the initial communication. Ask them if any of the people you intend to invite should be dropped from your list. If so, ask for the names of others who should be considered instead.

- Here is an example of an e-mail message that covers these points:

> From: David Benson
>
> To: Project Team Leads
>
> cc:
>
> Subject: Please mark your calendars—Conference Call January 15, 2004
>
> The purpose of this note is to let you know that there will be a conference call January 15th from 2:00–3:00 p.m. CST. The objective of the call is to discuss the upcoming reorganization—specifically how it relates to our teams.
>
> You will receive another note closer to the 15th with the dial-in information and a detailed agenda.
>
> Please make every effort to participate. If you believe that another team member should also attend, please contact me.
>
> **David Benson**
> **ABC Consulting Company**
> **Manufacturing Project Manager**
> **Direct dial: 1(555) 555-5555**

- Send the agenda, the call-in number, and any materials at least two days prior to the call. Remember, some people might be traveling and will be unable to check their e-mail every day.

- As appropriate, request that participants review topics/objectives for the call and suggest changes.

- If the announced conference call will focus on a complex issue or a topic of debate, you might consider contacting some of the key participants to provide a high-level overview of the call and get their buy-in and/or support of the topic. Use your best judgment here.

Conducting the Call

- Do not drink carbonated drinks prior to or during the call. You don't want to burp during a conference call. (Oops! I guess I should have told you that earlier.)
- Remember, conference calls are really meetings with "invisible" participants and co-speakers. Act and speak as if you are conducting a meeting.

  Be prepared to think on your feet. Sometimes conference calls do not go as well as planned (e.g., someone might get disconnected, you might have a bad connection, etc.). Just be prepared, and don't hesitate to get the operator back on the line.
- Speak clearly and confidently. Consider standing up; standing up as if you are presenting sometimes helps you project your voice with more clarity and authority. Standing also provides more freedom of movement so that you can be animated, which will be reflected in your voice.
- Welcome participants as they join the call. If the conference call involves more than 20 participants, consider having music playing until the conference starts.
- At the beginning of the conference call, provide an overview of what the call is about, the objective of the call, how the call will be structured, and what participants can expect.
- If the call is for a very large group and/or many of the participants have not participated in many conference calls, consider setting ground rules. If the participants are seasoned business professionals, ground rules will probably not be necessary—especially since the ground rules will probably be similar to those used in a meeting.

- Here are a few key ground rules to start you out. You might want to add a few of your own.

- Be an active participant and listener.
- Do not speak over one another.
- Show respect for your colleagues.

There are several others that could be listed. For a more complete list of ground rules, refer to Steve Kaye's book, *The Manager's Pocket Guide to Effective Meetings.*

- Here are a few additional ground rules more specific to conference calls:

- Provide your name before you speak, since people won't be able to see who is talking.
- If you must leave the call for any amount of time, alert the facilitator and the remaining participants, since they won't be able to see you leave the room.
- Do not eat or chew gum.
- Do not place the call on hold if your phone has "music hold."
- Place the phone on mute to minimize background noise.
- Keep the tone of the call positive.

- Take a roll call of the participants so that you can take the appropriate follow-up measures with those who were not present. If the number of participants is large, request an attendance report from the service providing the conference call.
- If the number is large and not everyone knows one another, consider taking a more specific roll call. State what group or function each participant is from as they are identified so that everyone has a general understanding of which groups are represented.

- If there are more than eight or nine participants, remind the participants that if they are on a speakerphone, they should use the mute button on their phone to minimize background noise.
- Be attentive to the audience. Remain focused on the objectives throughout the call.
- Be aware of the tone of your voice. Keep the call upbeat, even during difficult discussions.
- If you are the facilitator, be sure that your personal opinions or emotions do not inappropriately influence the participants, and vice versa.
- If you must use a conference call for a brainstorming session, ask participants to withhold judgment until the end of the session.
- *Never* use offensive language or comments, no matter how well you know the audience. It is unprofessional and might offend participants and presenters.
- Never put someone on the spot or embarrass them on a call.
- If you are using Q&A with a large group, ask each person to state his or her name before they ask a question. This will help the facilitator make the conference more personable and make follow-up less difficult. It will also help participants who wish to talk with that person later on about a point or question raised during the conference.
- Respect each individual by keeping the call on schedule.
- If a discussion begins and is not helpful for the majority of the participants, offer to continue the discussion off-line (and facilitate if necessary).
- If your phone system plays music when a call is placed on hold, do not place the call on hold at any time during the conference call.

- As in a meeting, everyone should participate. If you have not heard the opinion of one or more of the participants, ask them to share their perspective.
- Don't let one or a few people dominate the call.
- Verbally recognize when participants make contributions to the call. Perhaps the topic of the call is the new expense policy that was distributed last month, and several people are discouraged and feel that it's not working. If several participants express their concern and dissatisfaction and another participant asks to share a positive view of the policy and something that is working for him, recognize and encourage this kind of behavior for the remainder of the call. To do this, thank the participant for his perspective and ask open-ended questions during the discussion. Sometimes this helps other participants become more engaged, rather than disgruntled or bored.
- Be sure to cover all the topics on the agenda adequately.
- Provide humor, and personalize the call where appropriate. What would the world be without humor?
- If the call runs longer than an hour, check after an hour has passed to make sure that participants can accommodate the length of the call.
- If a few participants are in the room with you, use them as a gauge. Make eye contact to determine whether or not people understand what you are saying.
- At the end of the call, ask yourself (and the participants, if necessary) if the objective of the call was met, or if they need more time to discuss the topic at hand.
- Make sure you conclude the call with an overall summary. Provide next steps if necessary. If there is a to-do list, make sure everyone understands time frames and requirements for their tasks.

- Be sure to thank those who participated in the call. Let them know that you are open to feedback on how the call went, as well as suggestions for future calls.

Following the Call

- Send minutes from the conference call immediately following the call so that key issues discussed during the call are still fresh in your mind and in the minds of the participants.
- Be sure to provide a clear summary of the next steps to be taken.
- As appropriate, follow up with those individuals who were unable to participate in the conference.
- As appropriate, follow up with those individuals who struggled or had many questions.
- Make notes for future calls about what aspects of the call worked well and what can be improved.
- If you expect to have regularly scheduled calls with the same participants, follow up after the first couple of calls with a few of the participants to see if they have suggestions for future calls (e.g., logistics, format, content).
- Begin planning your next conference call!

As a Participant

- Prepare for the call just as you would a meeting. Review material ahead of time and list questions and discussion points.
- Give the call your undivided attention. Avoid the temptation to multitask. Okay, let me just spell this one

out: No reading those 196 e-mail messages that you've received today. *Yes, you know what I'm talking about!* Doing this could cause you more work in the end because you will miss critical components of what was discussed.

- If you have trouble hearing the facilitator or other participants, politely tell the facilitator. Don't be shy. There might be a problem with the phone line.
- Do not dial in using a cellular phone, unless the call is critical: Cellular phones sometimes cause bad reception during the conference call, and typically have a lot of background noise. If you must use a cellular phone, notify the facilitator ahead of time, if possible.
- If you are using a speakerphone, place your phone on mute.
- Avoid background noise, if possible. Don't shuffle papers, don't whisper to someone else, don't close desk drawers, etc.
- State your name before interjecting yourself into the discussion.
- Have respect for the other participants. If others are asking questions and you already know the answers, listen and be patient. (You might hear something new.)
- If clarification of a topic is needed and you feel that you can add value to the discussion, politely ask the facilitator if you can interject and provide your perspective.
- *Never* use offensive language or comments.
- If your phone system plays music when a call is placed on hold, do not place the call on hold at any time during the conference call.

- If you are with other participants in a room, avoid the temptation to converse intermittently. If you must say something, wait for a break and indicate to the others on the call that a point or question related to the topic has come up and that you must place them on hold or mute to discuss it with the other team members in the room. When you return to the call, be sure to report back what was discussed and any new ideas or resolutions.
- Do not chew gum if you are going to be speaking. If you must partake, please do not smack.
- Do not eat. (An exception to this is when the call takes place during your lunch hour and/or you just really need to eat because you've been on the phone all day and are famished.) If you must eat, please eat quietly. No one should hear you crunching on those wonderfully tasty chips.

5

Web Conferences and Videoconferences: Just Virtual Meetings

September 11, 2001, changed everything. Terrorist attacks, a fear of flying, and a sluggish economy pushed companies to seek ways to get the same work done quickly and efficiently without having face-to-face meetings. Collaborative technologies such as Web conferencing and videoconferencing are meeting the need by offering companies alternatives to travel. While these technologies are not meant to replace *all* travel, if they are used effectively, they can substantially reduce and/or postpone staff travel and related costs.

Web conferencing is one of the fastest-growing segments of the collaborative market. An estimated $500–600 million spent on Web conferencing in 2003 is expected to grow to almost $2 billion by 2006. Market growth for videoconferencing is expected to exceed that: Purchasing and leasing videoconferencing equipment and public room services is expected to reach or exceed $4 billion by 2006 (TIA March 7, 2003).

The substantial growth of these collaborative communication technologies in an otherwise shaky economy is an indication that companies are seeking a different way to bring employees together to get work accomplished and achieve their target results.

In this chapter, we'll provide a brief introduction to Web conferencing and videoconferencing, and follow up with tips on how to plan and manage these conferences and information about a few vendors who provide these services.

WEB CONFERENCES

Web conferencing is fairly new to the business world, and might seem a bit technologically scary. Don't let it frighten you. With today's economy, business professionals really need to look at different ways to conduct business—not only to reduce costs, but to keep people closer to home (which is where they want to be). Today's conferencing technology, combined with Web conferencing capabilities, creates the ability to almost re-create face-to-face meetings.

Web conferencing enables you to present PowerPoint slides, draw pictures, review tools and Web sites with your team members, play games, and much more. While it's true that the presenter controls what participants see on their screens, Web conferences are collaborative: They offer features such as online chat, whiteboards, user polls, and live video transmission. Sound complicated? Well, it's not. It's just as easy as participating in a conference call and accessing a Web site.

The reason I say "participating in a conference call *and* accessing a Web site" is because I prefer to use a conference call in addition to Web conferencing, instead of using the audio piece of the Web conference technology. The audio quality is usually better on the telephone (the quality is getting better over the Web every day, though, so one day this will no longer be true). The conference call technology also gives people a comfort zone, so using both technologies offers you a way to transition people to Web conferencing. As participants get more comfortable using Web conferencing and the quality of transmitting sound over the Internet improves, we will all be more likely to rely solely on the Web conferencing capabilities. It does keep costs down if you use only the Web conference, rather than use both conference call and Web conference.

Web conferences can also be an effective way to train employees. E-learning is a growing field, offering countless opportunities to train according to the need of the moment. Many training sessions today are held remotely: Participants sit in various locations and are merely linked by their laptops and a telephone. In these settings, participants dial in to a conference call or a Web conference to link with their other "classmates" at scheduled times during the training. This linked time gives participants the opportunity to discuss "live" what they have learned so far, and to share experiences, ask questions, etc.

VIDEOCONFERENCES

Do you remember when videoconferencing capabilities first came out? The concept was great. I don't know about you, though, but the technology side of it seemed much too complex for me. However, we are becoming more accepting of technology in general, and smarter about how to use it (whether we've wanted to or not). Videoconferencing is beginning to catch on.

> Total spending on the videoconferencing market grew 36.7 percent in 2002 to $2.05 billion, and is expected to grow 24.4 percent in 2003, reaching almost $2.6 billion. . . . In addition, the emergence of public videoconferencing rooms attracts infrequent users to participate in the market

> without incurring capital or staffing costs. Approximately 7,000 public videoconferencing rooms generated $600 million in revenues in 2002 (*TIA,* March 7, 2003).

With those growth estimates, I think it's clear that we are getting past videoconferencing's identity problems. However, I still don't think many companies view videoconferencing technology as a "must have."

Videoconferencing offers obvious advantages.

With videoconferencing, you can hold a meeting in two locations in the same city or many locations across countries *at the same time.* In some circumstances, the presenters might want to travel in order to be in the same city to facilitate a large conference call, but most participants will not need to travel unless their location does not have the appropriate equipment.

When it comes to enriching communication, what if the president of your company wants to announce to all 20,000 employees the possibility of a merger? Rather than breaking the news through a memo or voice-mail broadcast, he might want to deliver it personally and "meet" with everyone within the company. How can he do this without having one big, expensive meeting or several smaller meetings? Videoconferencing or some type of satellite broadcast immediately comes to mind.

Business leaders have begun to use videoconferencing or Web casting as a way to more personally announce significant organizational changes to all of their executives globally. A videoconference might not sound very interactive if you've never participated in one, but it can be! It can also be quite motivating to link up people who share the same goals. If it is done correctly, it can be almost as motivating and inspiring to the participants as a meeting.

Just as in a meeting, a videoconference enables speakers to use whiteboards, presentation slides, and body language to help get their point across. A good videoconference feels just like a good meeting. The participants receive handouts, follow an agenda, hear from various speakers, get to ask questions, and have scheduled breaks (if necessary).

Now that you are excited about the possibilities that videoconferencing provides, I want to encourage you to take a step back and think about the other side of the coin: When *not* to use videoconferencing. Videoconferencing is great for the example mentioned above in which a CEO wants to "meet" with the employees to make a big announcement, but it's important to recognize that we should not use videoconferencing just to use it. Chapter 4 discussed the possibility of combining a conference call with a technical solution using an application such as Netmeeting. Videoconferencing might be less critical in these circumstances.

Let me provide you with an example in which videoconferencing was probably not the best solution. I was recently talking with a friend of mine who participated in his first videoconference. He said that he was quite distracted by "seeing" the person he was talking to. The distraction wasn't with the individual, but rather with the technical component of "seeing" the person. Specifically, he stated that the voice did not match up with the mouth movements. The individual spoke with a heavy accent, so my friend was also trying to focus on understanding what was being said. This was extremely frustrating to him. When I asked why a videoconference was chosen over a conference call, his answer was, "I don't know." He further stated that there was no reason that he needed to "see" the person. In this situation, it sounds like an expensive videoconference took place when a conference call would have been a better choice. The moral of that story is just to be sure that a videoconference is necessary. A conference call or Web conference can be cheaper, easier to set up, and a better answer for what you are trying to accomplish.

The purpose of this chapter, however, is to provide you with what you need to do once you have determined that a Web conference or videoconference is the best solution for you as a speaker, a facilitator, or a participant to communicate more effectively and get the most out of the experience. The purpose is not to provide you with all the technical things that you need to think about. The content of this chapter also assumes that you have the technical equipment, the connections, and the space required to conduct a videoconference or Web conference. For more information on equipment and connection needs, you should contact the telecommunications or IT manager of your company or one of the vendors listed below. If you do not have someone like that as a resource and you will be the one responsible for setting up everything, there are several sites and companies on the Internet that can help you.

Some helpful vendors and resources include:

Books

- *Videoconferencing: The Whole Picture,* by Trout Bayard
- *Cybermeeting: How to Link People and Technology in Your Organization,* by James L. Creighton and James W. R. Adams
- *We've Got to Start Meeting Like This: A Primer for Videoconferencing,* by Edward A. Daly

Vendors

- www. messagebank.com
  This vendor provides Web conferencing capabilities, as well as 24/7 audio-conferencing numbers.
- www.videobridging.com
  videobridging services
  videoconferencing tips
- www.conferencecallUSA.com
  videoconferencing services
  conference calls
  Internet conferencing services

- www.globalvideoconf.com
  video equipment rental
  video equipment leasing
  videoconference event staging
  conference room design and installation
  public room rentals
- www.placeware.com

Virtual Communications Toolbox

- http:// www.presentplus.com
  This resource provides information on Web conferencing and Web casting.

Organizing a videoconference or Web conference is a little more challenging than setting up a conference call. But don't let that stop you. Consider hiring a professional facilitator to help you.

Aside from the technical components, prepare for a videoconference or Web conference much like you prepare for a meeting. Here are a few things to consider when preparing for and conducting a videoconference or Web conference:

Preparation

- Document the objective and expected outcomes of your message to make sure that a videoconference or Web conference is the appropriate forum for your message.
- Determine the audience.
- Once you have determined that a videoconference or Web conference is in order, begin working through the technical aspects and logistics of setting up the conference and reserving the appropriate space. (Web conferences might not require space, since individuals can participate via laptop.)

When you are ready to reserve space, be sure to select a room that is away from noisy areas (e.g., elevator banks, mailrooms, etc.). This is not a big consideration when you are setting up a face-to-face meeting, but it is very important for a videoconference.

- When you are determining the exact time for the event, consider the various time zones of the participants. If you can't seem to find one time that accommodates everybody, consider offering a rebroadcast to some locations.
- If multiple locations are involved, identify an individual within each location who can be a "host" for the videoconference. This person can help reserve space, help with communication, make sure handouts are ready, etc. More importantly, this person can be on the ground in that location to make sure everything goes smoothly and help with issue resolution. This step is not essential when you are holding a Web conference, but having people from the same location gathered together reduces costs and helps with "teaming."
- Send a memo to the individuals you are inviting. State the objectives of the event, as well as the date and time so that they can put it on their calendars as soon as possible. Let them know that the details and necessary materials for the conference will follow separately.
- Determine which materials will be needed for participants and speakers, such as handouts, flipcharts, markers, an LCD panel, etc.
- Arrange for necessary equipment to be placed in the room that has been reserved for the conference.
- Begin preparing materials for the participants. Remember to send participants handouts and reference

materials ahead of time so that they have them in time for the videoconference and can look them over.

- Consider giving your presentation a timed trial run.
- Plan to be at the meeting room 30 minutes ahead of schedule to make sure that everything is set up correctly and that materials are in place.
- Test the equipment to make sure everything is in working order.

Considerations for the Presenter

- Whether it is a face-to-face meeting or a videoconference, do not wear distracting patterns or colors. This is even more important to remember if you are a presenter at a videoconference. Certain colors and patterns tend to react adversely to light and distort the image that appears on the screen.
- If there will be a large number of participants, consider soliciting questions in advance.
- Welcome callers as they join. With a large conference, you can use a screen with a message (e.g., "Welcome to the Merger Meeting. We'll begin shortly.") and background music.
- Present yourself just as you would in a meeting. Do not let the camera intimidate or distract you.

Considerations for the Participant

- Spend time preparing and reviewing materials. Come with a list of questions and/or discussion points.
- Listen attentively.

- During the conference, jot down questions or discussion points until there is an opportunity for you to speak.
- When you have an opportunity to speak, be sure to state your name and location.
- Act just as you would in a meeting. Do not let the camera distract or intimidate you. Let yourself get absorbed into the content of the meeting.
- Also, be aware of what you wear for the conference. If you think that you might speak in front of the camera, do not wear strong colors or vivid patterns.

For more specific tips, contact the vendor who is providing the service.

6

E-mail, Voice Mail, Conference Call, Web Conference, or Videoconference—Which one should I use for my message?

Throughout this book, we've talked about the importance of sending your message using the appropriate vehicle (e-mail, voice mail, conference call, Web conference, or videoconference). By now, you might be asking yourself how you can be sure you have selected the best way to deliver your message. The vehicle itself is less important than the need for communication, but it is still a critical factor to consider.

This chapter provides guidelines that will help you determine the best option *for you* in a certain situation. Remember, there is no right or wrong decision. In the end, you will need to decide what is best for your particular situation. If you are communicating a simple, informal message to co-workers, you probably don't have to worry too much about how you do it. However, the method becomes more important when you are dealing with complex, detailed messages or communicating with various levels and groups within an organization.

A detailed list and notes of things to consider as you choose the vehicle for your message is provided below. A quick reference guide follows at the end.

Remember: These guidelines are not meant to be all-inclusive or used as strictly black-and-white rules. Just consider these ideas when you are unsure which vehicle to use for your communication.

Consider using e-mail when:

- Your message provides a lot of statistics, facts, numbers, and/or analytical data. This type of information is better shared in written form, to allow the receiver to reference and analyze the data.
- You need to share documents or you seek review of documents.
- You follow up from a conference call, Web conference, or videoconference with the action items/next steps that were outlined during the conference.
- You need the audience to review the contents of the note, and respond accordingly.
- The individuals with whom you would like to communicate are fairly knowledgeable about the subject at hand, and only limited discussion is needed.
- You, as the sender, need a dated/time-stamped record that the note was sent and a copy for future reference.
- You want to give individuals the opportunity to respond directly to you in writing.
- You want to introduce an idea or a topic prior to additional communication (e.g., a conference call).

- You are stuck on a conference call or you are in a meeting and must get a message immediately to an individual or group of individuals.

  I have used e-mail when I am on one conference call and am running late for another. In this situation, I send a message to the facilitator and/or participants (depending on the audience) via e-mail to let them know that I am still on another conference call and will be with them as soon as possible. You can also send a quick instant message instead if the individual is already online. (Please remember my earlier advice, though, that you really should try to avoid the temptation to multitask if you are participating in a conference call.)

- You know that the individual you are trying to reach is in a meeting or on a conference call and cannot check for voice mail messages, but is probably hooked up to his or her computer and is checking e-mail or instant messages in between topics or on breaks. In this case, use the subject line of the message to tell the individual exactly what you need so that they will understand the sense of immediacy.

  Many of my colleagues and team members are spread out across the United States, and we all participate in a lot of conference calls. As soon as we hear the "ding" for new mail, we'll glance to see what the subject is to determine whether or not it is urgent. Try putting something like this in the subject line to prompt quick return messages: "Help!!! I need to talk to you this morning, or the world as we know it will end." Of course, you will need to use your best judgment about putting humor into an urgent business request. When I used this line, it was *extremely* urgent that I talk with the individual I was trying to reach. It worked. As soon as she was finished with her conference call, she called me. (She was my boss.) Keep in mind, however, that sometimes this will not work. Try your best NOT to multitask on conference calls, and remember that the individual you are trying to reach might be so involved in his call that he won't even notice that he has new mail. Given the electronic world that we live in, however, the odds are good that this will work.

Consider using voice mail when:

- You want the audience to receive a message with a personal touch (through the sound of your voice).
- Your message is extremely urgent and you cannot reach the person directly on the telephone.
- You need to offer an explanation or you need to "talk" through an idea that might be difficult to express in a written note.
- The content of your message is extremely brief and you do not need a record that it was sent.

Consider sending an instant message when:

- Your matter is extremely urgent.
- You need to interrupt the employee from a task they are doing or about to do (e.g., you have a number to give them before they submit their budget).
- The content of your message is extremely brief and you do not need a record of it.
- You are on the same conference call as your teammate (only in different locations) and you want to provide them with some additional context around a subject on the call.

- You are a supervisor and want to provide some "just in time" suggestions to an employee who is presenting on a conference call. Wait until there is a breaking point before doing this. Otherwise, the employee will stumble in their presentation as they try to read your instant message. It works better if you both agree ahead of time to use IM. I used IM during a very complicated and somewhat political discussion to tell a team member what he needed to say to get through the conversation, almost word for word.

Consider scheduling a conference call or Web conference when:

- You need an open discussion around a specific topic, such as a new policy or a team discussion about the status of a project.
- You need to review with the audience various documents or tools (such as a new database, forecasted projections for your team) and get their feedback in an open forum.
- You need to brainstorm an idea.
- You want to train team members on a new application.
- You need to get your virtual team together for a "pep rally."
- You need status updates from your team. For this, consider setting up standing weekly conference calls with your team.

Consider holding a videoconference when:

- You need to send a significant leadership message (e.g., reorganization, new leadership).

- A virtual interview is needed. A videoconference virtual interview might pose some challenges (primarily equipment-related), but if you can overcome them, consider a virtual interview as a good alternative to flying a candidate in for a face-to-face interview.

  You are likely to resort to a videoconference virtual interview only when someone within your organization has already met the candidate or if the candidate is a rehire into or has transferred over from another part of your organization. It would be more of a confirming interview to "meet" the person.

  It's important to be aware that virtual interviews are not yet common; the candidate might therefore be more nervous than usual.

  One interviewing option used a lot in large organizations with virtual management and teams is to have someone in HR or recruiting conduct an initial face-to-face screening interview and then have the immediate supervisor and appropriate others conduct follow-up and confirming interviews by telephone. This works out well as long as the individual conducting the initial face-to-face screening interview is familiar with the skills and professional characteristics the hiring group is looking for in a candidate.

  If you choose to conduct a telephone or videoconference interview, here are a few things to consider:

  - Send any introductory or discussion material out ahead of the call so that people can prepare.

  - Try to make the candidate feel at ease. She might not be used to participating in these types of interviews. You can do this by making "friendly" statements or small talk at the beginning of the call.

  - Before you get started with the interview, ask the candidate if she has ever participated in a virtual interview before. This will help you understand her level of comfort and you can manage the interview accordingly.

- Clearly state how the interview will work on the phone—discuss any expectations you have and what the candidate can expect.
- Give the candidate time to react verbally.
- Listen for verbal cues. It is extremely important that you recognize and pay attention to these signals, since you will not be able to see the individual's body language. Even if it's a videoconference, most equipment is not able to really pick up subtle body language.
- Close the interview much like you would a face-to-face interview, using some type of "verbal handshake" and an explanation of next steps.

The reference matrix that follows can help you decide which communication vehicle is most appropriate for your situation. It is not meant to be all-inclusive or to provide you with the "right" solution. It is up to you to choose what's appropriate, based on your needs.

Quick Reference Matrix

Item to Communicate/ Communication Circumstance	Conference call	IM	E-mail	Voice mail	Video-conference	Web conference
Minutes from a conference call or meeting.			X			
Documents for review and response.			X			
Documents sent that need to be discussed.	X					X
A topic that is very familiar to the recipient(s). Little explanation is needed.			X			
The sender needs a date and/or time-stamped records.			X			
The sender wants to provide the recipient(s) with an opportunity to respond to you directly in writing.			X			

continued

You are on a conference call and need to get an urgent message to someone. Remember: You should not multitask during conference calls. This would be an urgent situation and an exception.		X	X			
The individual you need to reach is in a meeting, but you know he/she has a laptop and might be checking e-mail messages on breaks.		X	X			
You want the audience to receive a more-personal message.					X	
You need to brainstorm an idea with other virtual team members.	X					
You need to discuss status updates for a project or projects.	X				X	
You need to conduct a virtual interview.					X	

continued

Quick Reference Matrix *(continued)*

Item to Communicate/ Communication Circumstance	Conference call	IM	E-mail	Voice mail	Video-conference	Web conference
You need to share a significant organizational message.	X				X	
The message you want to send is very brief and you do not need a record that the message was sent.		X		X		
You cannot reach the person directly, but you need to send an urgent message with some details. You want to be able to pass along the details to ensure that he/she gets it correctly.		X		X		
You need to kick-off a new virtual team.	X				X	X
You want to train team members on a new application	X					X

Note: You will often need to use a combination of these tools in order to communicate effectively.

7

Legal Considerations

ABUSE, MISUSE, AND PREVENTATIVE ACTION

> Write as if you are being published, speak as if you are being recorded, and act as if you are being videotaped.[1]

The use of electronic communication devices and functions such as e-mail, voice mail, camcorders, tape recorders, and cell phones are exposing companies and individuals to civil and criminal liability at alarming rates. In today's litigious society,

---

[1]From *Preventative Law 101* by Eartha Jean Johnson (president of LegalWATCH). (For more information on LegalWATCH, access *www.legalwatch.com* or call their corporate headquarters at 713-864-9997.)

it is imperative that we act and communicate as if a record is being made of our every move or utterance, irrespective of our surroundings. The platinum rule to live by is to write as if you are being published, speak as if you are being recorded, and act as if you are being videotaped. You never know who is watching, listening, or recording or what single act or communication will be the key piece of evidence in a legal proceeding against you or the company for which you work.

Cutting edge technology is constantly being introduced. We can now communicate in ways and at levels that were unimaginable just a few years ago. We can communicate with people around the world and even in space, looking them in the eye as if we were in the same room. No matter how physically far away a person is, he is as near as his or her palm pilot, telephone, cell phone, fax, or computer. And whether we are in a restaurant or at the movies or are jogging in a park, driving down a highway, or visiting another country, we can be reached within seconds.

Unfortunately, the convenience, efficiency, and economical advantages associated with electronic technology are exposing companies and their employees to increasing liability. Because it is fast and convenient and considered informal, people tend to be casual in their electronic communication.

We begin Chapter 7 with an overview of the kinds of electronic media available and the sensitivities associated with their use. We conclude the chapter with a summary of recommended practices that can help minimize or prevent liability associated with the use or misuse of electronic communication.

E-MAIL

E-mail, by far, poses the single greatest threat to corporate viability. In recent years, the outcome of trial after trial has often come down to a simple e-mail. In fact, one could say that the fate of Arthur Andersen was sealed by a single e-mail. In the lawsuit brought by the Justice Department against one of the world's largest accounting firms for the destruction of documents, jurors delivering the fatal verdict said afterward that the shredding of documents was not a factor in the verdict—but a single e-mail was. In a patent-infringement case brought against a major U.S. company, one e-mail influenced the jurors to return a $98 million judgment against the company.

People often forward documents to others without viewing the attachments. Forwarding privileged, confidential, or proprietary information can be enough to waive a privilege. For instance, if your attorney tells you something that is covered by the attorney-client privilege and you forward the communication to another person outside of the privilege, you can waive the privilege. Likewise, if you send proprietary information to a third party and do not have a confidentiality agreement in place, that third party might be forced to later disclose the proprietary information. To protect proprietary information, one must show that safety mechanisms were in place to protect the proprietary nature of the communication.

People send e-mails to anyone and everyone, oftentimes without thinking about the consequences of the communication being made public. Sometimes people send copies of e-mails to others in an effort to impress or wow them with their knowledge. Doing this can be beneficial, but it should be done with caution. When we copy others, we have created another source of information about a particular subject matter who can be called upon to give a deposition or to testify at trial.

Employees often copy top-level executives on e-mail in an effort to protect or cover themselves. The problem with copying senior-level executives is that they often do not read their e-mail. If a senior executive is asked in court about a particular e-mail and has to admit that he or she did not read the e-mail, it is likely to place that executive in a negative and embarrassing light. Executives simply do not have the time to read all their e-mail messages. To make matters worse, the executive might read an e-mail and do nothing about it, assuming that someone else is handling the matter. The executive is then placed in an even worse situation.

At one time or another, most of us have either inadvertently sent an e-mail communication to an unintended party or received one that was not intended for us. Depending upon the content of the communication, this can be problematic, especially when the communication is of a confidential, personal, or sensitive nature. A high-ranking official at the Air Force Academy inadvertently sent sensitive information intended for a select group of commanders to 4,400 cadets. The officer said he simply hit the wrong button.

Laptops

The convenience of laptops further compounds the risks associated with e-mail and other electronic forms of communication. We can take them virtually everywhere we go to access the world at the click of a button. We thus increase the number of e-mails we send and that increases the number of potentially problematic situations we might find ourselves dealing with, not to mention the paper trails we create that detail every minute account of any given situation.

There are privacy issues associated with laptops. We see individuals working on their laptops on airplanes, in restaurants, and in hotel lobbies, apparently unconcerned about who might be looking over their shoulder. This can lead to a breach of confidentiality and can potentially reveal sensitive and proprietary company information.

Conference Calls

Conferencing capabilities have made it possible for us to speak with hundreds of people around the world at one time without leaving our office, our bedroom, or even our bed. People use conference calls in lieu of face-to-face meetings.

They can indeed save money and time by reducing traveling expenses and the time individuals must be away from the office. This convenience, however, can come at a great cost if precautionary measures are not put in place.

Employees often create liability for their employer. A manager once called to terminate an employee and used an audioconference feature to put another manager on the phone to witness the termination. When no one answered the employee's phone, the manager left a message. When the manager finished dictating the message, both managers continued to talk about the employee's overall conduct and performance deficiencies. One of the managers commented that she had witnessed similar behavior from other individuals of the same ethnicity as this particular employee and suggested that it might be a personality trait common to that particular ethnic group. As they continued to talk about the employee, an electronic recording interrupted their conversation with, "You have fifteen seconds to complete your message." Needless to say, the managers were in a state of shock. They thought they had disconnected from the employee's answering machine, only to find that their conversation had been recorded. They said things that not only incriminated the company, but also gave the employee ammunition to support a case for discrimination. This is a classic example of employees creating liability; the company had legitimate grounds for termination, but the company ended up paying far more to handle the aftermath than they would have otherwise.

SPEAKERPHONES

Whenever you are speaking with someone via a speakerphone, it is important to immediately tell the other person that they are on speakerphone. People tend to think that telephone conversations are private and that no one else is listening. Failing to warn others at the beginning of the conversation can lead to embarrassing moments for both parties and expose the company and the offending employee to liability.

Tape Recordings

There are times when it seems wise to tape record a telephone conversation. Proceed with extreme caution: Several states (Virginia and California, for example) require that you inform the other person that you are taping the conversation. In some states, only one party to a conversation needs to know that the conversation is being taped, and the party taping the conversation can be that one party. Make sure you know the laws of the particular state so that you don't open up yourself or your company to a criminal offense.

Some companies have policies prohibiting the taping of a conversation and even go as far as stating that taping a conversation can be grounds for dismissal. A company with such a policy might be able to terminate an employee for not following the policy, but courts are not bound by a company's policy, Therefore, an employee could use that recording in a court of law. The best advice is to always communicate as if you are being recorded. Assume you are being recorded, even if you are told that you are not. A person can lie to you and still be able to use the recording in a court of law.

Surveillance Cameras

Surveillance cameras are everywhere. The average citizen is captured on various forms of surveillance an average of 200 times per day. In some cities there are cameras attached to light poles to record motor vehicle violations. When we enter a grocery store, a bank, an office complex, an airport parking garage, a college campus, a discount store, a restaurant, a parking lot, a school, or a house of worship, our actions are usually captured on some form of surveillance. We can never be too careful; someone is always watching our moves—maybe even at work.

In summary, here are a few tips to consider:

- If you need to continue to speak with an individual after the conference call is over, disconnect the call and call the person back. This is especially important if the subsequent communication is of a sensitive, confidential, or personal nature. This ensures that only the individual you want to speak with is privy to the communication.
- Remember: Whatever you transmit from your computer can be recorded. Electronic transmittals expose individuals and companies to liability. Most people think that because the transmittals are not saved to the hard drive, there will be no record of the communication. Not so. There are computer software programs that take pictures of your computer screen at specified intervals. Therefore, I advise you not to type anything at work that you do not want made public.
- Do not use work computers for personal matters.
- Be careful who you send a copy of information to. Make sure the person has a "need to know."
- Do not forward e-mail messages without reading attachments.
- Always be alert and aware of your surroundings and the information you might be exposing to the public. Be careful what you say in person or on a cell phone in a public place, and be careful about any information on your laptop screen. The same goes for chat room discussions.
- And finally, remember this advice: Write as if you are being published, speak as if you are being recorded, and act as if you are being videotaped.

8 Conclusion

It's All in How You Use It

Electronic communication isn't the perfect solution for everyone's particular circumstance, but it is here to stay. The ability to use e-mail, instant message, voice mail, Web conferencing, and conference calls effectively and appropriately is essential to career success. Communication is a critical part of working on a virtual team, and it is just as critical in building professional relationships: Electronic communication tools make it possible to work more closely with clients, teams, and individuals quickly and efficiently, regardless of the physical location.

This chapter will focus on how to use the tools that we have to communicate with virtual teams; work effectively from a home office; and build and manage solid professional relationships, even if those relationships must be managed across cities, states, or countries.

Communicating with Virtual Teams

A virtual team is made up of individuals who are not co-located but who share a common goal, objective, or vision. Good communication is essential to building and maintaining successful virtual teams. Virtual teams are becoming more and more a part of the workforce as companies become global and more professionals telecommute. The focus of this book is not on how to build virtual teams, but on the communication tools that are essential in business today, and virtual teams are now a key component of the business world. Given that, and the fact that sometimes it helps to see tips and tools applied in real-life situations, let's spend a few minutes looking at how you can communicate more effectively using the tools we've discussed, and how you can communicate more effectively with your virtual team.

E-mail

E-mail is essential to keeping your teams linked together. It facilitates the sharing of information, and sharing information is a key component in the success of a team, especially a virtual team. With e-mail, you can create and distribute status reports and scheduling details in a timely manner.

- **Weekly status reports and/or newsletters**

  Managing a virtual team is not the same as managing an on-site team. When you manage a virtual team, you cannot be a micro-manager. It's impossible. You must focus more on what the team is doing and what they are getting accomplished, rather than on how they are accomplishing it. Weekly status reports help you do that: You can provide a clear understanding of what your team is working on and the issues they might need your help with. If individual status reports are created by each member of your team, you can consolidate them and create one team report each week or month for distribution to each person by e-mail.* This helps each team member know what the others are working on, thus creating a common ground, which is really important for virtual teams.

- Here is a sample template for a weekly or monthly status report:

February Accomplishments

Task/Project/Goal	Comments
ABC Budget Report	Analyzed budget and provided variance report.
New York Training Sessions	Confirmed presenter assignments for February and early March sessions.

March Targets/Plans

Task/Project/Goal	Comments

*Before you distribute the report, first review it for information that might not be appropriate to share with the rest of the team. Consider talking this over with your team members to make sure they are comfortable with this.

Future Initiatives

Task/Project/Goal	Comments

Management Issues

Task/Project/Goal	Comments
Need additional resources to complete the 1-2-3 Project review and analysis	Project analysis is due to management by April 15.

- **An "Items to focus on this week" memo**

  If you manage a team in multiple locations, consider sending an "Items to focus on this week" memo at the start of each new week or month (whatever approach best fits your work objectives). Each staff member has a long list of items to complete each week, but it's a good idea to keep people focused on the big deadlines and/or goals they need to meet for their team.

  - Here is an example of an "Items to focus on this month" memo.

> We are at the start of a new month. I wanted to take the time to thank all of you for the hard work last month. We met and exceeded our goal for new-client work by 10%. Way to go!!! After reviewing the status reports each of you submitted, I realize that we still have a lot to accomplish. Therefore, I wanted to take a moment to provide you with a list of things that our team should focus on this month.

continued

> This is a guideline to help you plan your month. Please let me know if you have any questions or concerns.
>
> **Personnel Administration:**
>
> - Complete updates to personnel records for your team by May 15th (e.g., terminations, leaves of absence, transfers).
> - Finalize all performance evaluations for the individuals you supervise by May 29th.
>
> **Sales Forecasts:**
>
> - Complete all new-client paperwork by the end of the day on Thursday, May 20th, and submit it to your Region Lead.
> - Submit sales forecasts and expense estimates for the next month to your Region Lead by close of business day on May 8th.

Here are other kinds of communication that can be shared through e-mail:

- **Birthday/anniversary announcements**
- **Tools/reference documents and other resources/databases**
- **Department budgets and forecasts**
- **Team contact lists/organizational charts**
- **Project and other organizational documentation**
- **An idea mailbox**

  An idea mailbox enables your team to send ideas as they think of them to you or the entire team.

  With an idea box, you can capture ideas and retain them for future reference. Maybe the idea is not right for this time and place, but you might need that idea down the road as circumstances change.

- **Team member profiles**

  Need a team-building exercise? Consider having each team member create a personal/professional profile as a Word or PowerPoint document that can be shared through e-mail. The profile might be purely business, personal, or a combination of the two.

I prefer a profile that is business and personal, sharing things such as experience, length of time with the organization, various roles the individual has played within the organization, current skills/strengths to offer the team (e.g., budgeting or presentation skills) and fun things such as their favorite food, favorite city, outside interests, etc. Listing skills and strengths is helpful because team members can use one another as resources when they need help in a specific area. If I've never done budgeting before and I need to develop a budget for my department, for example, I can contact another member of my team who has noted budgeting on her profile as a strength.

You could request that each team member include a scanned photograph as well. This might be especially important if you feel that the members of the team won't get an opportunity to meet in person. This will help them put a face with a name and help them build relationships with other team members.

Whatever you decide to do for your team, ask your HR manager to review it to be sure that the components of the profile are acceptable from an employee-relations perspective. You'll need to be careful to avoid questions about age, gender, race, marital status, etc.

An example of a profile is shown on the following page. Create your own questions. You might consider asking each of your team members to complete this profile, get together on a conference call, and introduce themselves to the team using the information in their profile of themselves.

> **Name**
> **Title/Role within organization**
> **Work phone number/e-mail address**
> **Start date within the organization**
> **Birthday (no year, please)**
>
> Primary responsibilities and customer base:
>
> Key projects that I'm currently involved in:
>
> Projects I'm interested in learning more about and/or would like to be involved with:
>
> If you ever need help with______________, I might be a resource.
>
> Things I like to do in my personal time:
>
> My favorite food/movie/city/color . . .
>
> If I could spend one day alone with someone (living or deceased), it would be with_____________________________.

Voice Mail

Voice mail is a great communication tool for teams because it delivers information quickly, with minimal technical risk (it's doubtful that the phones will be down), and it has an informal connection. Voice mail can also be fun. Unlike e-mail, it lends itself to personalized messages, since your own voice will be heard.

You can use voice mail to communicate many things. Here are a few specific suggestions that will help your virtual team stay in touch and up-to-speed. You can:

- Create "pep rally" broadcast messages.
- Send promotions or birth announcements to your team.
- Send birthday/anniversary celebration messages to your team.

- Use voice mail for information that needs to be broadly distributed. "Broadcast" or "radio" messages can be created that provide regular updates to your team.
- Use voice mail to offer a detailed explanation of complex documents sent through e-mail.
- Use voice mail for a quick announcement (e.g., to set up a conference call).
- Send a message to each person assigned to your location who is staffed out-of-town (but who gets to come home for the weekend), highlighting weekend activities. This will help those individuals still feel part of a "home" team.

Conference Calls, Web Conferences, and Videoconferences

Conference calls and/or videoconferences are essential if you want to effectively communicate with a virtual team.

Once teams become comfortable with conference calls and videoconferences, they will begin to see them for what they are: just virtual meetings. These technologies provide us with opportunities to share perspectives and opinions and discuss issues "live."

Conference calls, Web conferences, and/or videoconferences can also be used for team building. How is team building done on the phone? Most team building is indeed done in person, but it is possible to develop teams using a combination of conferencing technologies. Here are a few examples:

- Remember the earlier section on creating profiles of the members of a team? Consider using a conference call to have each individual present his or her profile to the team. This is an extremely valuable exercise in the beginning stages of a team. When new members join a team, it can be used as a great introduction activity. (However, if you've already passed that "introduction" window, it can still be a useful tool.)

Other activities suitable for conference calls or videoconferences are:

- **Virtual meetings/parties**

  Virtual parties, virtual meetings . . . Whatever you want to call them, they are a lot of fun and can help you meet a defined business objective or share project status, financial plans, etc. Let me explain. You can have a guest speaker at a virtual meeting and even a theme, such as celebrating team success (i.e., meeting sales goals). Imagine yourself sitting in a room alone, getting ready to participate in a team conference call. You are looking at a box or an envelope your team leader sent in advance of the call. The package reads, "Do not open this package until you receive instructions on Tuesday's conference call." You are intrigued, and you can barely stand the suspense. *What's in that package? Why can't you open it?* Well, the reason you can't open it is because part of the virtual meeting is having everyone open the package at the same time. You dial in to the conference number provided, and you hear the song, "You're Simply the Best." Just as you start to wonder if you have dialed in to the right conference call, the leader of the conference call announces, "The ABC Project Team Conference Call will begin shortly. Please do not open the package you received until you are instructed to do so. If you did not receive a package, please push # and then '1' on your phone so that the operator can alert us."

(Wow! This is getting exciting. Are you with me? Do you visualize what is happening? Let's say you are a team of 10 individuals spread across the United States. Members are each sitting in a different city, yet all are wondering what is happening.)

Now, back to the call. The leader comes on and introduces the call with great enthusiasm, and instructs everyone to open their box. Everyone is in listen-only mode while they open their box. You can't hear them, but the anticipation is building.

What's in the box? Well, I will stop there. That depends on what you want to do. You can have party hats, balloons, sparkling grape juice, and a plastic champagne glass—whatever you wish. The key is to make the call fun and help the team feel linked.

So, we've got the fun covered. What about the "linked" part? People should feel linked, but just to be sure, you might want to ask them to all do something at the same time, such as "On the count of 3, we will all blow up a balloon and tape it to the wall." Just choose something.

I've got you thinking, don't I? To take it one step further, ask team members to gather in a conference room (if you have multiple team members in one location) to participate in the call. You can have the room decorated with party stuff, have a cake brought in, and so on. The possibilities are endless.

- **Sing-alongs**

Okay. I'm sure you are a bit surprised by the title of this section. It's not often that you can pick up a business book and read about sing-alongs, is it? Well, here you have it . . . building and managing virtual teams in a new millennium.

Let me explain what I mean about sing-alongs. I believe that being spontaneous is sometimes the best way for a team-building activity to come about (given that the spontaneity is appropriate). I recently facilitated a conference call during a really rough period of time for my team. We were a three-week-old team in the middle of a reorganization and in new roles that we knew little about (though many of us had worked together before). There was a lot more work than many of us anticipated. Toward the end of the call, I picked up on a general sense of heaviness among members of the team. At that moment, I felt a song in my heart (and I am *not* musically inclined!). I told the team that I understood that there was a lot of work to do, but that we'd pull through it (typical leadership message, but I truly meant it). I told them that I felt a song coming on. Well, several people started laughing, which made me feel better. Just hearing their laughter made me realize that we'd be okay. I told everyone that I am not musically inclined and didn't feel that they needed to be tortured, so I would spare them from my singing. However, I shared with them the title of the song and some of the key lyrics. Several individuals commented, and I could tell they felt my sincerity.

Well, as I was preparing for the next conference call, I decided that I needed to get my hands on that song. As I struggled to locate the music and the lyrics, one of the participants from the original call contacted me to let me know that *she* had found the song. She repeated some of the lyrics that I had referred to before. I was so excited that I immediately voiced my intention to use it for our next call. That's when I started secretly planning a sing-along.

In preparation for our next conference call, I put together the agenda with all the information we needed to cover. The last topic was "Be Prepared." In that section, I attached a sound byte and lyrics. I don't think many people figured out why I did that, but when we got to that section of the agenda and I explained what I had planned, several people laughed and some groaned. I had gone too far to turn back at that point, so I continued. Guess what happened? When the sound byte started, people sang! And, from what I could hear, it sounded like everyone participated (no way to tell, though)! Sounds a bit silly, doesn't it? Nevertheless, I received a lot of positive feedback. My biggest concern now is trying to come up with something fun, different, and spontaneous for the *next* conference call. . . .

A word of caution: You might not want to try this on the first couple of conference calls with a new team. It is just too risky; they'll think that you're crazy (they don't know you yet). Wait until the team feels a little more comfortable with one another first. You might want to save this idea and use it just about the time that things get a bit boring.

Another word of caution: If your team is small (under 10), a sing-along might just be a little too uncomfortable.

- **Game shows**

  Game shows are fun and educational. You're probably thinking, *What is she talking about?* I've actually used game shows with my teams, and it has been a lot of fun. You can revolve the questions around the organization. Here are some good categories:

  - Leadership/organizational structure
  - Mission/vision/core values
  - Organizational history (e.g., when the company was founded, by whom, etc.)
  - Support (e.g., mailroom, Technical Support, etc.)

You can tally points for each conference call and give away small prizes. You can create various categories for prizes (e.g., best guess, most excited voice, etc.). You can also keep the tally going for several calls to keep up the excitement. If the game is successful, you'll have team members just waiting for the next call. Obviously, you should play this game at the end of scheduled topics. On second thought, why not add some fun by having one to two questions at the beginning of a call (after the concept catches on)?

- **Lottery**

The lottery game is a quick way to break the ice on a long conference call; to get calls off to a fresh start; or to energize participants during a long or tedious call. The game is played much like it sounds: whoever "holds" the number that is drawn, wins. To play this game, send everyone on your team a number or have someone else on your team do this if it seems more fair. Then, at the desired time, draw a number and there you go.

I receive extremely positive feedback whenever I use this lottery game.

- **Guess who this is**

I often use this game when I realize that my team will not be meeting for quite some time. The more members you have on your team, the better it works.

A day or so before a regularly scheduled weekly conference call, each team member receives by U.S. mail an unidentified photograph. Attached to the photograph is a Post-it that says, "Save for our Wednesday call." The team should already have created profiles without the photos, so the people in the photographs will look like strangers. Tell the team that the photos are of their virtual teammates, and that each member has to guess the identity of the team member in their photo. (*Note*: You can do this electronically, but I find the element of surprise in receiving the photo in the mail to be half the fun.)

In building virtual teams, it is extremely important to make sure that team members understand their role, expectations of performance, and the team's purpose and mission so that they can work toward common goals. This chapter focuses on ways to build and maintain your virtual team, but when a team is first forming, you should set up a true face-to-face meeting. However, if a face-to-face meeting is not possible, be creative in developing some of the key components of a solid team (e.g., trust, strong relationships, common purpose, feeling of community). You'll also need to regularly spend some one-on-one time on the phone with each team member (say, once a week) to make sure everything is working well. You will find that as you think through these ideas and get more experience, new ideas will come.

I realize that some of the suggestions that I have provided in this chapter are unconventional. However, that's sometimes the best way to create a climate of trust among team members. As I have said, building and managing a virtual team is different than managing a team that is co-located. A virtual team needs to feel connected in other ways; sharing the same silly experiences has proven successful for me, and it might work for you.

See how creative *you* can become in communicating with your virtual teams. Experiment and find out what works and what doesn't. The objective is to create an environment where it almost feels like participants are physically near one another. Videoconferencing can really help with the feeling of "being there," but some of the ideas we've described for conference calls get you pretty close.

Working from Your Home Office

Technology gives us the ability to work virtually anywhere: on airplanes, in hotels, from our home office—all without skipping a beat. Keep in mind one thing: It is very important to maintain a professional atmosphere and attitude for our clients, regardless of where we are working. Let me explain. Whether you are working at home in your office, downtown in your office, or in a hotel in another city, everything about your environment should be invisible to your clients and co-workers: no dogs barking, no babies crying, and no washing machine noises in the background. (You don't have these things in your downtown office.) Background noise can be extremely distracting to those talking with you on the telephone.

There are other things to keep in mind as you work from a home office.

If you work consistently from home, even if it is only one day a week, I strongly recommend that you do the following:

- Get a second phone line or a DSL line. This will allow you to be on the phone and on your computer at the same time.
- If your business line can be forwarded from your primary office to an alternate number, have it forwarded to your home office. This will make it easier for your customers and co-workers to reach you.
- Be sure to answer the phone using at least your full name. (You might also consider using your company name; it helps create the environment of a business office.) Your clients shouldn't have to ask or wonder with whom they are speaking. I can't tell you how many times I've contacted someone at their home office, only to have them answer the phone, "Hello." I always have to ask who is speaking.

- Get a phone answering service or professional voice mailbox to record incoming messages. This way, if you are on another call, you can be certain that the individual contacting you will receive a professional message versus your personal home answering machine. If you choose to use a voice mailbox, be sure to include your name and your company name in the introduction.
  - Example:

> "Hi. You've reached John Benson with JohnArt Gallery in Houston. I am currently on the phone or away from my desk. Please leave your name and phone number, and I will return your call as soon as possible."

- Turn your call-waiting off each time you place a telephone call. Contact your phone service provider to find out how you can do this. On my home telephone, I select *70 on the keypad. Turning off your call-waiting helps ensure that you will not receive any annoying beeps while you are talking. If you have a voice mail service or phone-answering service, your call will be picked up immediately if you are on the phone and a message will be recorded.

- Be certain to check your messages as soon as you finish a phone call. Check them frequently throughout the day so that you can return calls promptly.

Building and Maintaining Effective, Professional Relationships in a Digital Age

I often hear people say that our digital age has depersonalized business and removed much of the "relationship" aspect of business. Relationships continue to be a critical component of successful business, and effective communication is the key to building effective relationships.

You might find yourself struggling with aspects of e-communication, but don't let the digital age stand in your way of building effective, long-standing business relationships. It's all in how you use the tools that you have.

Oftentimes, businesses reduce travel costs in an effort to reduce overall operating expenses. If your client is in another city, state, or country, you might never get the opportunity to meet your key customers, have lunch with your favorite client, or take to dinner that client you are afraid you'll lose to your competitor. However, there are a few easy things that you can do to build relationships in any environment. I'm sure you can come up with many others of your own. In fact, I'd recommend developing a brief "relationship" plan for each of your key clients or customer groups. This will help ensure that you focus on building and maintaining those relationships. Try these suggestions:

- Trust. It is a critical component of any relationship.
- To help build trust, stay in touch. Frequent, open, and meaningful communication is critical to building solid relationships.
- Always communicate the truth, even if it's painful.
- Follow up on time and as promised.
- Add a personal touch to e-mail, voice mail, and conference calls. How can you do that? Easy. One way is to generate sincere "small talk." For example:
  - Ask about your colleague's or client's weekend or kids.
  - Send a card when he or she is sick or has had a death in the family.

  - Recognize key life events.
  - Talk about sports or an interest you both share.

- Monitor tone, even in stressful situations. Tone of voice or tone of words in e-mail is extremely important, because it's all your client has to go on. They can't see your facial expressions or body movements.

  Stress often tests relationships, but remaining calm when presented with stress will build confidence and trust.

- Give the other person your full attention when on the telephone. Do not multitask by checking e-mail messages (unless, of course, they refer to something they sent you and ask you to open it). Do not talk to another person in the room while you are still "on" with them. Doesn't that just burn you up when the person to whom you are talking says something totally out of context? You ask, "What?" and they answer, "Oh no, not you. I was talking to ________."

- Treat the other person with respect.

- Treat them as they want to be treated. If you don't know how they want to be treated, find out.

- Know their communication preferences (e-mail, voice mail, or telephone call), and use that form of communication whenever possible.

- Treat them as if you might lose their business today, regardless of how long they've been your client.

- Always give them the benefit of the doubt when something goes wrong. Remember, most of us are reasonable, rational people.

- Assume nothing.

- If your company is experiencing change that will impact your client(s), help them understand the change early on—not just the details, but the bigger picture as well. Early, clear communication helps people embrace change quicker than if it catches them by surprise.

I hope you have found this book helpful. It has been fun sharing my ideas with you. Good luck, and remember: Communicate unto others as you wish others to communicate with you.

Additional Reading Resources

1. *Better, Faster Email: Getting the Most Out of Email,* by Joan Tunstall
2. *Cybermeeting: How to Link People and Technology in Your Organization,* by James L. Creighton and James W. R. Adams
3. *Mastering Virtual Teams: Strategies, Tools, and Techniques That Succeed,* by Deborah L. Duarte and Nancy Tennant Snyder
4. *New Communication Environments: From Everyday to Virtual,* by Giuseppe Mantovani
5. *The Manager's Pocket Guide to Effective Meetings,* by Steve Kaye
6. *The Virtual Edge: Embracing Technology for Distributed Project Team Success,* by Margery Mayer
7. *Tools for Virtual Teams: A Team Fitness Companion,* by Meg Hartzler and Jane E. Henry, Ph.D.
8. *Videoconferencing: The Whole Picture,* by Trout Bayard
9. *Virtual Leadership: Secrets from the Round Table for the Multi-Site Manager,* by Jaclyn "Knights of the Tele-Round Table" Kostner
10. *Virtual Teams: Reaching across Space, Time, and Organizations with Technology,* by Jessica Lipnack and Jeffrey Stamps
11. *We've Got to Start Meeting Like This: A Primer for Videoconferencing,* by Edward A. Daly

Helpful Web Resources

Vendors

- www. messagebank.com
  This vendor provides Web conferencing capabilities, as well as 24/7 audio-conferencing numbers.
- www.videobridging.com
  videobridging services
  videoconferencing tips
- www.conferencecallUSA.com
  videoconferencing services
  conference calls
  Internet conferencing services
- www.globalvideoconf.com
  video equipment rental
  video equipment leasing
  videoconference event staging
  conference room design and installation
  public room rentals
- www.placeware.com

Virtual Communications Toolbox

- http:// www.presentplus.com
  This resource provides information on Web conferencing and Web casting.

INDEX